Successful Farming

All Around The Farm®

75TH ANNIVERSARY

The best ideas by farmers, for farmers that you can put to use today

Seventy-five years ago, in October 1929 to be precise, Successful Farming® magazine introduced a new feature that was not only for farmers but by farmers. Now here was something different. It was a simple concept. Let farmers share their ideas for solutions to everyday problems around the farm with the rest of the magazine's readers.

Perhaps no other group of individuals had a reputation for being more ingenious than farmers. And they had plenty of ideas.

Necessity is the mother of invention

This rural "ingeniousness" had a lot to do with the shortage of, and sometimes even nonexistence, of equipment and materials for use on the many far-flung and isolated farms throughout the United States. Farmers often had no alternative but to make do. The old saw "necessity is the mother of invention" was more reality than cliché on the farm.

Farmers were, in fact, quite proud of their problem-solving abilities. But with the exception of neighbors, their inexpensive solutions and inventions generally went unrecognized. More often than not, they were eventually forgotten.

Our most popular page

That is until the editors of *Successful Farming* had the bright idea to solicit ideas from farmers and publish a page of them in every issue of the magazine. The section was called All Around the Farm®, and not surprisingly, it was an immediate hit. *All Around the Farm* was soon the most popular and most read section in the magazine, a distinction it continues to hold to this day, 75 years later.

Practical and inexpensive ideas

In those 75 years, *Successful Farming* has received an untold number of ideas and published over 6,500 of them. The ideas you'll read in this book represent the best farmer submissions of the past four years.

Farmers have always appreciated the ideas of other farmers, because they know they'll be practical and inexpensive. Even more important, they're tried and proven to work. We think you'll also enjoy the pure and simple ingenuity of many of these ideas. They get the job done, and whether it's on the farm or around the home, that's still the bottom line.

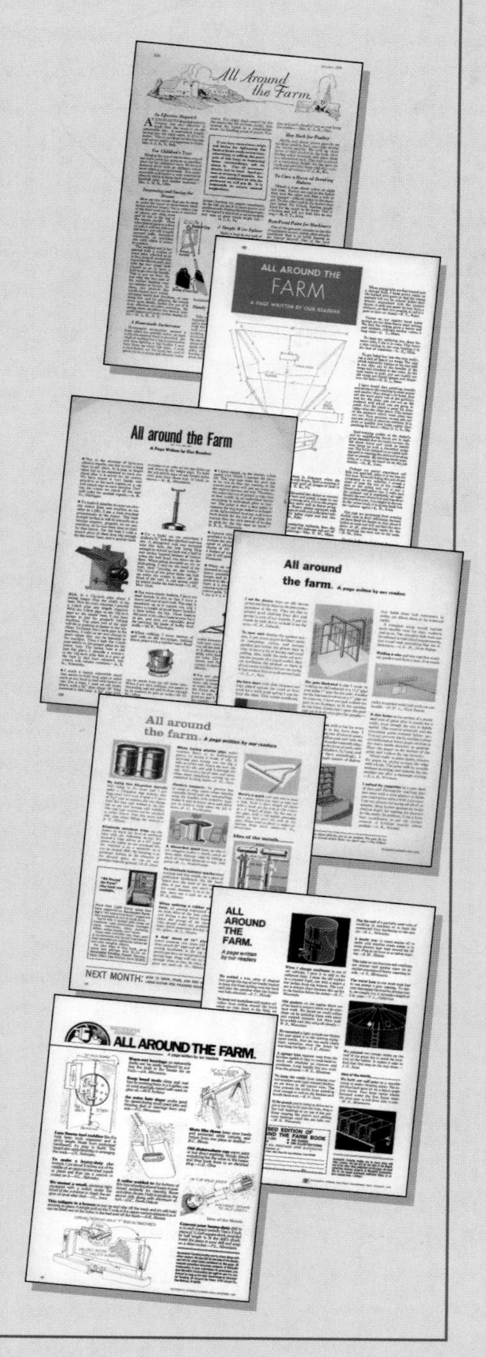

Successful Farming

All Around The Farm®

Twelfth Edition • Fall 2004

In Memorium -- E.T. MEREDITH III (1933-2003)

Portions of this book have been previously published in *Successful Farming* magazine.

Successful Farming is published monthly by Meredith Corporation. If you have comments or questions about the material in this book, please write to *Successful Farming* magazine, Meredith Corporation, 1716 Locust St., Des Moines, IA 50309-3023, or e-mail brenda.torsky@meredith.com, or call 800/678-5752.

ISBN: 0-696-22433-X (tradepaper)
ISBN: 0-696-22366-X (spiral)

Contents

4

Drafty door repair

Our shop floor is so uneven, the seal on the over- ...

Bale grabber

BRACKET

To improve the holding power of the bale grabber mounted on our tractor's front end loader, I welded a bracket to the hydraulically moveable arm and bolted on a 14-inch tire and rim. The rubber sidewalls greatly reduce slippage on wet, snowy, or ice-covered plastic wraps.

–G.G., Massachusetts

68

Auxiliary ATV toolbox

An 8-inch-diameter piece of PVC piping makes a cheap and easily accessible extra toolbox for our ATV.

We glued a cap on one end and a handle on the cap on the outer end. We mounted it under the rear fender, using pipe clamps, It is out of the way but easy to get to.

–D.H., Nebraska

Spotlight on-the-go

I welded the female end of an air chuck to a flat piece of iron and screwed the male end into our spotlight. Then I bolted that piece of iron onto our ATV. Now we can quickly attach the spotlight to the ATV and take it with us. Then it detaches just as easily from its place on the ATV. Most importantly, it's there when we need it.

–K.R., South Dakota

Cargo rack enhancement

I covered the front and rear cargo racks on my ATV with ⅞(inside dimension)x½-inch-thick pipe insulation. This keeps the ATV from getting scratched when I'm carrying equipment. Also helps prevent slipping, so I can get by with fewer straps.

–S.B., Ohio

Auger-moving ATV

Using our ATV makes it simple to push out the auger's swingout and to pull it back out from under our semi trailers. We made a tow bar out of two seed corn signposts. The posts have holes running the entire length, which make it adjustable. This certainly does save back strain.

–S.S., Ohio

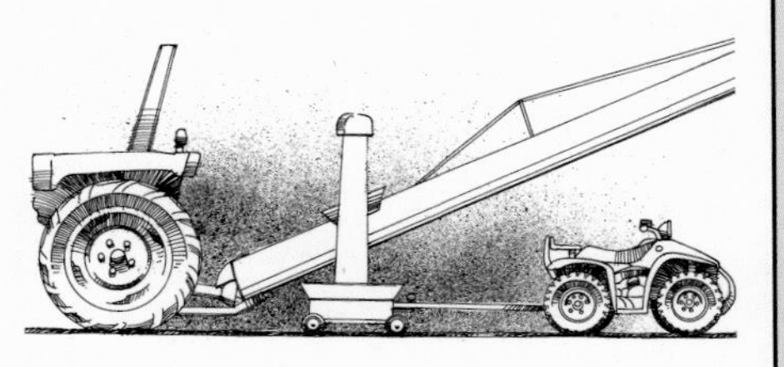

Handheld GPS computer on a four-wheeler

With Pocket Dlog, your ATV is the mouse and the world your drawing pad. Data log borders, creeks, drains, selective soil sampling, etc. Good background images are bare soil, Veris, and aerial imagery.

–Posted by ahinkson

Angled blade ends

We designed ends for our blade that angle to accommodate the blade's angle. They're easy to put on and take off, too. We can use the blade with only one end angled at a time since they angle individually. Straighten or angle each end just by removing clips. I push snow, manure, and level our gravel lane.

–W.H., Kansas

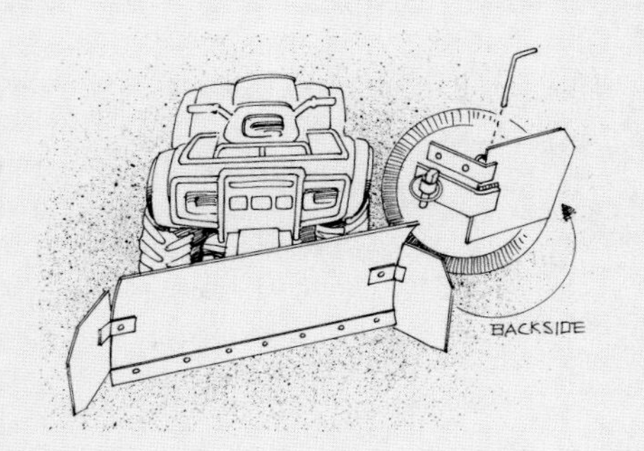

Note: To purchase kit with blade ends and brackets to attach them to an ATV blade, call Jackson Welding at 785-242-4300.

ATV towing

I use this hitch to tow my four-wheeler behind my equipment. It's made of scrap iron but could also be constructed of telescoping pipe. Towing saves time, work, and extra steps since now I always have a ride home for lunch or repairs when there's a breakdown.

– H.M., Montana

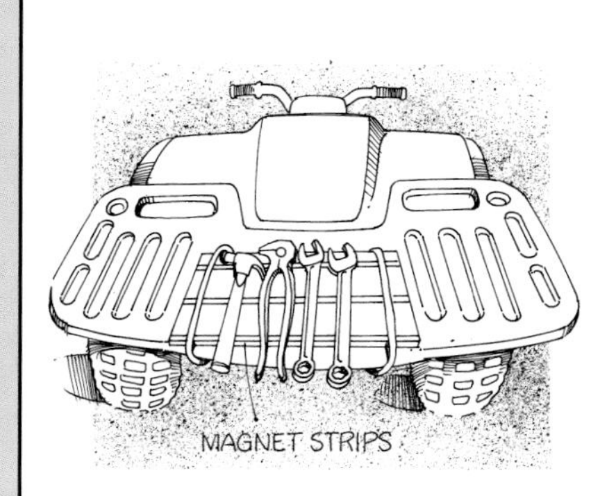

ATV toolholder

An old grinder mixer magnet mounts to my four-wheeler to hold my fencing tools and other hand tools. A piece of inner tube underneath keeps it from scratching the ATV. It goes on and comes off easily with two tarp straps. Now I don't have to worry about losing my tools out in the field.

–J.S., South Dakota

Fender flare extensions

Here's an inexpensive alternative to fender flare extensions for four-wheelers. Trim the top rim off plastic cattle protein tubs with tin snips. Then cut the width of plastic you desire for your flares and screw onto fenders or existing fender flares.

–C.S., Oregon

Utility trailer with dump bed

This 3x4-foot-long trailer stands 24 inches high. I constructed it from an old metal hopper. The trailer features a dump bed. The tongue is made out of thick-wall, 2½-inch square tubing and has a 15-inch drop hitch. It has 14-inch wheels that I took from the rear end of a front-wheel-drive vehicle. The trailer also has a 15-inch clearance underneath so it won't be too close to the ground. I pull it around with my ATV.

–G.M., Missouri

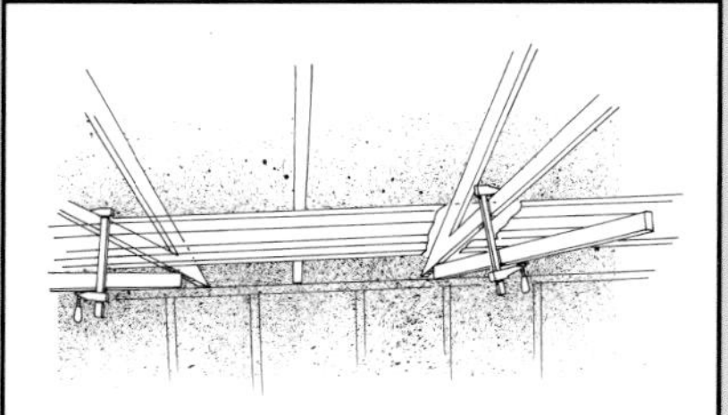

Ceiling clamps

These holders were very helpful when we put up our new metal shop ceiling. They are made from 8-foot 2x4s, two ratchet wood clamps, and two sets of fork guides. There are two sets of fork guides on each clamp that straddle the ceiling joists for stability.

With the metal held securely in place, hands are free to run screws.

–H.B., Pennsylvania

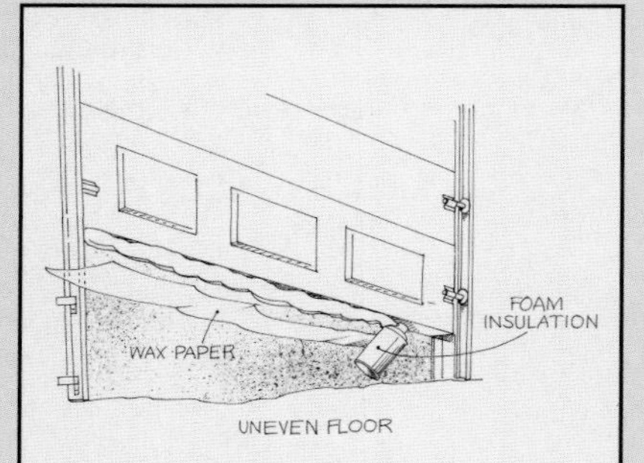

Drafty door repair

Our shop floor is so uneven, the seal on the overhead door would not work. I applied a bead of expanding foam insulation to the seal then covered it with wax paper. We closed the door, and the foam expanded to fill every gap. When dry, I removed the wax paper and trimmed the ragged edges.

–C.O., Wisconsin

Vapor barrier for metal shed roof

Happy with the insulation in a 40x70-foot shed we keep hay and machinery in, I'm using the same method in a 48x60-foot we're putting up. Put in ¼-inch-thick styrofoam with a plastic cover and stretch it tight, then tin goes on top of that. The styrofoam comes in 4x50-foot lengths.

–Posted by Infutures

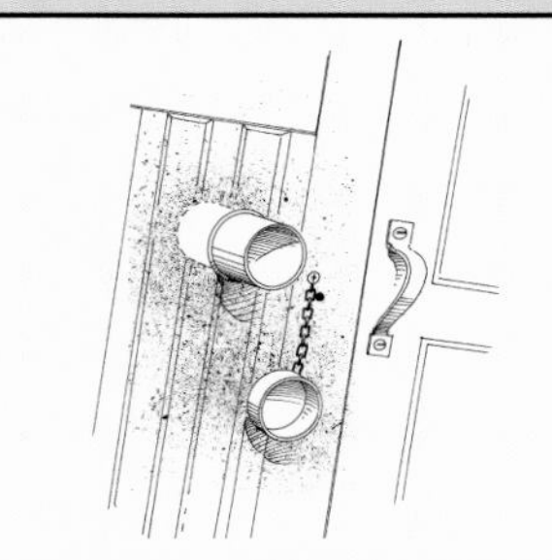

Fly-free milk building

This length of PVC pipe that I installed through the wall near the door of my milk building is a perfect way to keep flies out in the summertime and warm air in during the winter. The tank truck's hose fits right through. This way, the door stays closed while the milk is pumping.

–G.W., South Dakota

Movable freezer strips

We mounted freezer strips on a track for the scrape alley in our free stall barn. This makes it easy to remove them for cleaning or for summertime. The freezer strips are mounted between a 2x6 and a 2x4. The 2x6 is attached to the rollers.

–D.M., Wisconsin

Cheaper floor keeps the dust down

Instead of concrete, I put much less expensive crushed stone on my shop floor. There is 2-inch crushed rock (also called sewer rock) spread 2½ to 3 inches deep over the dirt. It needs no sweeping or cleaning. Reasonable amounts of tracked-in dirt will disappear within the stone.

–Posted by Robert

Insulation for sliding shop doors

We used 3½ inches of styrofoam and a plastic vapor barrier on our doors. Total R value is about 19. The added weight isn't enough to harm the track or rollers. Side walls have about R-17; ceiling is about R-35. Close the doors on warm spring days and you'll have a cool shop until late July.

–Posted by j42p

Obstruction-free doorway

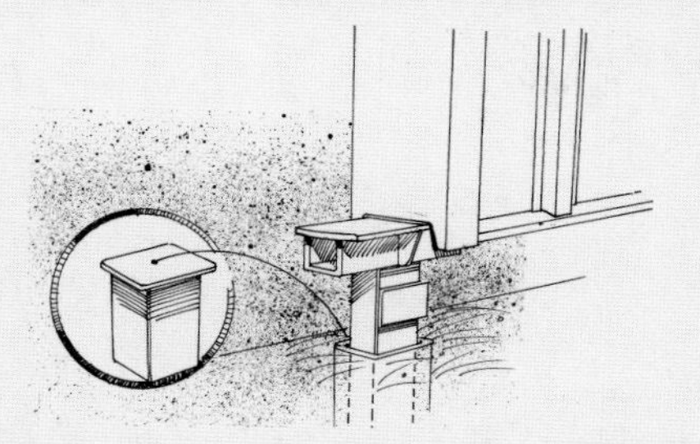

Weary of dodging the anchor for our big double doors, I dug it up and put a 3-foot length of 2½-inch-square tubing in concrete at ground level. Then I welded the door bracket to a piece of 2-inch-square tubing that slides down into the tubing in the ground. Now I don't have to worry about hitting the bracket with a tractor tire of a heavy, loaded wagon.

–M.R., New York

Keep doorknobs from freezing

We have an exposed, exterior door on our garage. I slide a 2-inch trailer ball cover over the doorknob and tip it up to keep rain and snow out. This prevents freezing because no water can set in.

–K.Z., Illinois

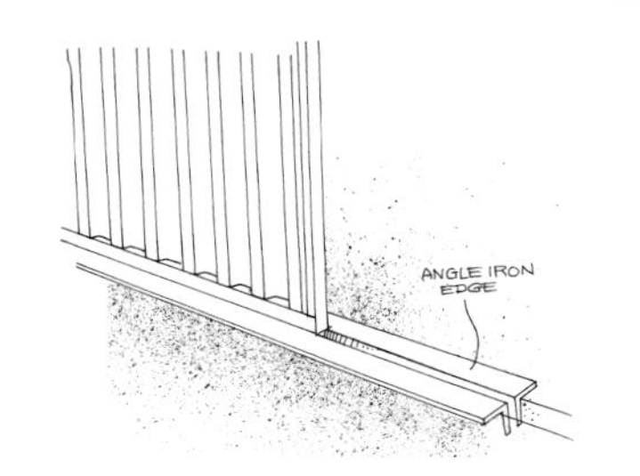

Angle iron guard

We added heavy metal angle iron to the edges of the cement where the shed door closes. It protects the edges from breaking off or chipping and keeps the wind from blowing the door off the hinges. The approach is poured so it catches about 2½ inches on the bottom of the door.

–K.W., Illinois

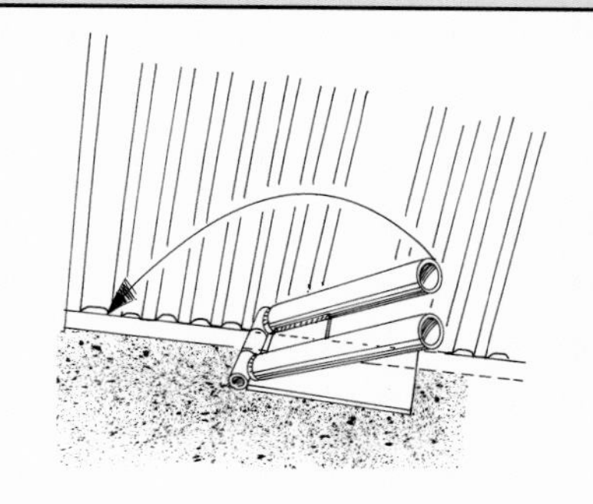

Extra door guard

Using a piece of ½-inch pipe as a hinge point, I made a fork out of heavy wall 2-inch pipe that swings up and stops at a 45° angle to hold both doors in place. When the doors are open, the fork lays down flat. A little loose sand keeps it from getting stuck down in the winter.

–D.C., Wisconsin

Save/throw away plan for major shed cleaning

When it's time for my annual cleanup, I use two empty pickups backed up to the door, a loader bucket, and a dump truck. The pickups take tires, parts, etc. to one of two smaller sheds. The junk goes right from the bucket into the dump truck that takes it to the dump.

–Posted by happyhusband

Walk-through door stays in place

We can sandwich our swing-out, walk-in door between two 12-inch lengths of rod slid into two pieces of conduit embedded in the cement about 5 inches apart.

–K.W., Illinois

Water shortage prevention

We have an outdoor Christmas bulb in a 120-volt socket wired to our well pump pressure switch. It comes on when water is running. The socket is mounted to a visible spot outside our barn.

–C.T., New York

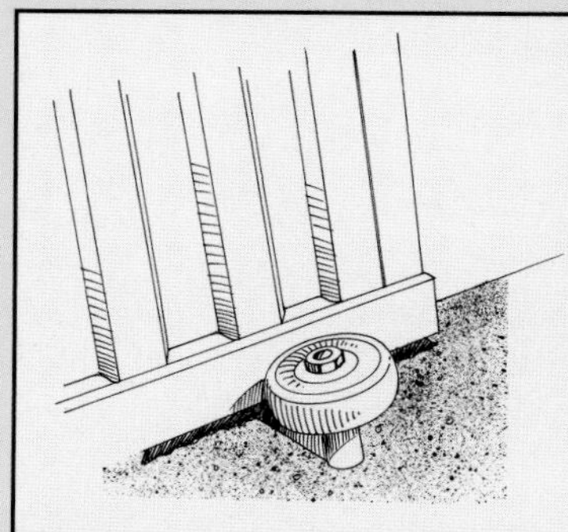

Replacement rollers

We've had problems with the plastic rollers on our machine shed doors getting brittle, then cracking and breaking. The solution was in a pair of outgrown in-line skates! The hard rubber wheels and bearings will hold up in all weather conditions for many years.

–L.W., Minnesota

50-gallon drums will protect seed

We've sold seed for over 15 years and used both plastic and steel barrels for rodent-free storage for 14 of the 15 years. With three 50-gallon drums, you can stack a 60-bag pallet off the floor that rats and mice will never get. If the drums are evenly spaced, the pile will never go down.

–Posted by cyfarm

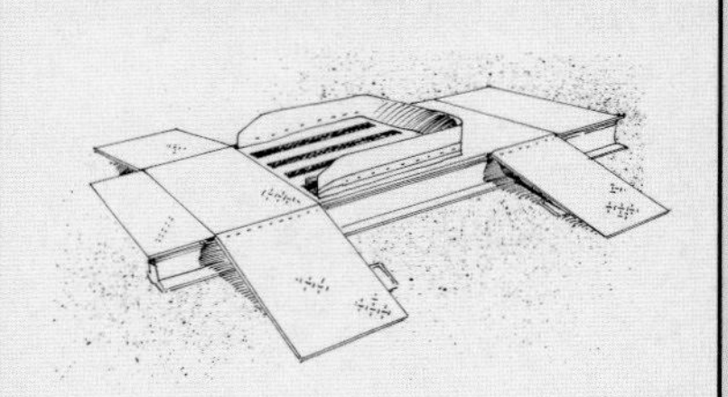

Sideboards for dump pit

We hated seeing grain splattered around our portable pit, so we extended rubber flaps around it. Anything with a low clearance like a hopper bottom semi can rub the rubber flaps with no fear of damage to the semitrailer or the portable pit.

–E.T., Illinois

Rid rodents from alfalfa fields

Using a spade to level off the newest mound, I dig down about 1 foot with a ⅜-inch rod. I insert 2 feet of 1-inch propane hose hooked to the anhydrous tank and turn on the gas. Works well for me.

–V.G., Nebraska

Ready reference

Laminating population and vacuum charts and taping them on the planter boxes near the transmission saves a lot of time during planting. You don't have to look through the owner's manual every time you need to reference the charts.

–Posted by glennw

Cord holder

This holder keeps the electrical cord held back out of the sweep auger while cleaning out grain bins. Weld a ⅜-inch bolt on one end of an 18-inch-long ⅜-inch rod and fasten it to the motor bracket. Bend the other end into a circle.

–K.B., Illinois

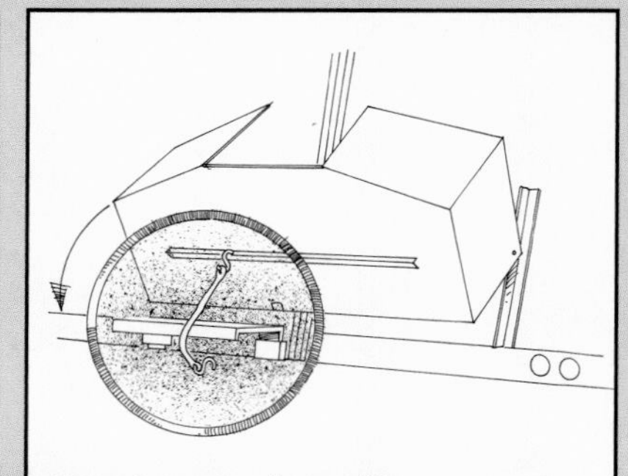

Easier unloading

We alternate between unloading in an auger at our grain bins and going to the grain terminal at harvest. To make taking off and putting on the grain chute easier, I made a metal holder for support so one person can do the job.

It slides into a slot under the doors and stores under the truck seat. I also use a tarp strap to help keep it in place for added security.

–R.H., Indiana

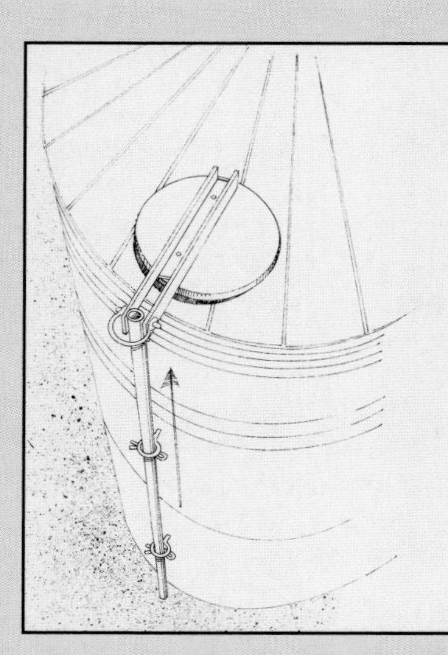

Open grain bin lid from down on the ground

I removed about 6 inches from the end of a small piece of aluminum channel and fastened it to the lid of my older grain bin. Then I connected a piece of gray pipe to it and ran it down the length of the bin with two I-bolts.

A cotter key fits in the two holes I drilled at the bottom and holds it open or shut. I have a removable clip for the entrance doors.

–E.C., Ohio

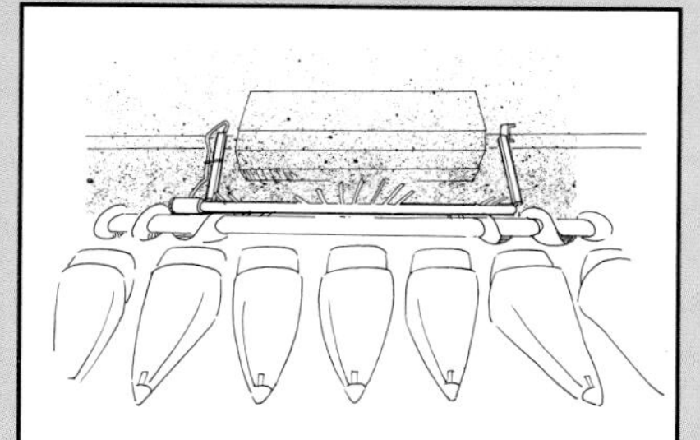

Clearing cornstalks

To combat the stalk buildup in the center of our 12-row corn head, we mounted a hydraulic motor on one end of a small silage beater. We speed up or slow down the motor-driven beater with the reel speed button in the cab. Now we can go 4 mph instead of barely moving.

–M.K., Minnesota

Drive-over dump pit

This 11x25-foot concrete drive-over dump pit was designed for unloading our corn to the dryer bin. The auger's swing hopper is positioned right below the steel grate. My son and I built it ourselves, and I'm sure it is the best $1,600 I have ever spent.

–R.A., Minnesota

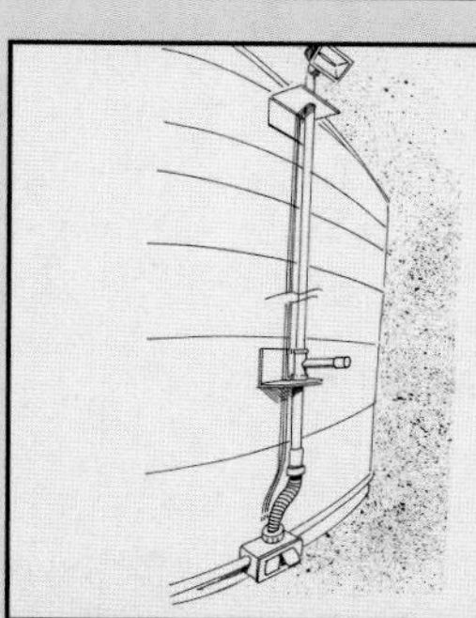

Directional lighting

This light is mounted on conduit, which rotates in two brackets with a T-handle. Change the light's direction depending on the truck's position when loading and unloading at night. Flexible tubing on the bottom keeps the rain out.

–K.W., Illinois

Controlled burn supplies

This rack, built from scrap pipe and angle iron, holds all the tools I use for controlled burning (firepot, water backpack, various hand tools, etc.) It's convenient to make one tractor hookup to all these well-used tools. Controlled burns are very popular down here in plantation country.

–R.T., Georgia

Fast, easy filling

I use a calf bottle to put graphite powder into planter boxes. But I made the hole at the end larger so the powder flows out.

–A.K., Illinois

Cone-shape soil probe penetrates the ground better

We built an inexpensive, user-friendly soil test probe using an ax handle and a cone-shape piece of sharpened metal. The sharpened bottom end helps get it into the soil and stepping on top of the cone helps as well. Since the cone is tapered out at the top, the cone will hold several samples before dumping.

–K.W., Illinois

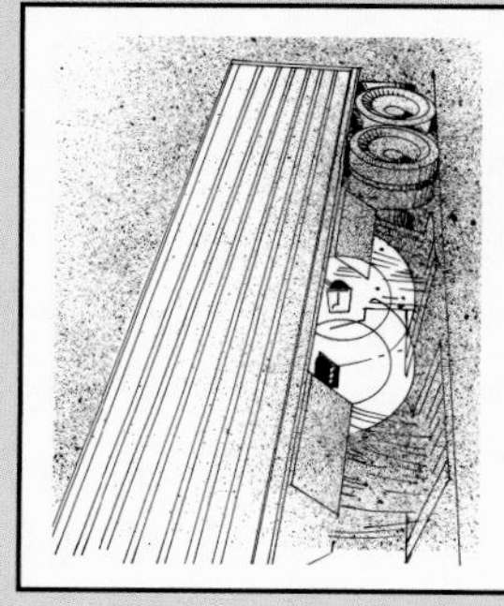

Well-lit hoppers

Two lights, one on the front hopper's backside and one on the rear hopper's front side, shine on the opposite hopper bottom to help monitor grain flow into the side swing auger of my PTO-driven grain auger. A toggle switch located near the lights makes it easy to reach.

–P.D. North Dakota

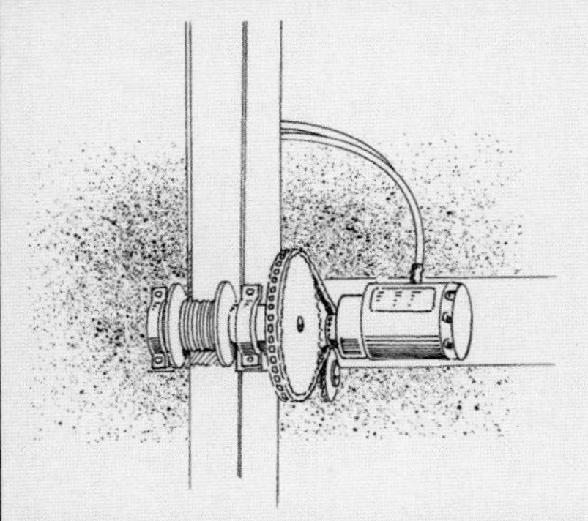

Control from the cab

We removed the hand crank winch from our grinder and replaced it with a spool made of heavy steel that is run by a slow-moving hydraulic motor. A large reduction gear between the motor and winch is connected by chain to turn it even more slowly.

–J.K., South Dakota

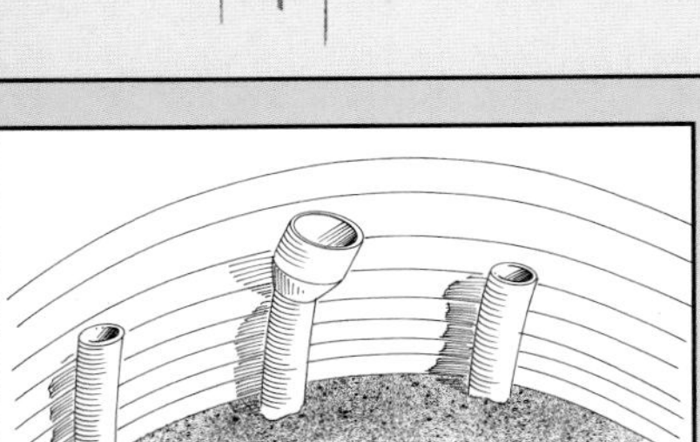

Aerator fans for bins

We were always having spoiled corn around the edges of the bin. We put 6-inch plastic drain tiles about every 4 feet around the edges of the bin. We keep it up high enough to avoid the sweep auger on the bottom. By moving a small fan around the bin, we can avoid the expense of running the big fan in the bin.

–L.S., Minnesota

Easier way to empty grain bins

A leaf blower is convenient for moving grain left behind by sweep augers in our grain bins. Both corn and soybeans are blown easily to the gathering side of the sweep auger. I don't need to shovel anymore.

–D.S., Illinois

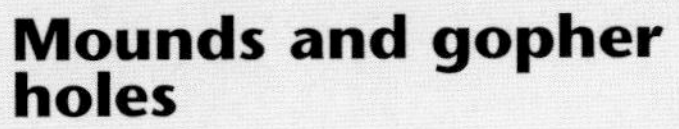

Mounds and gopher holes

Carry a gallon jug of field rye seed and sprinkle any obstacles with the seed. Next spring the tall rye marks the obstacles so the operator can prevent damage.

–B.D., Michigan

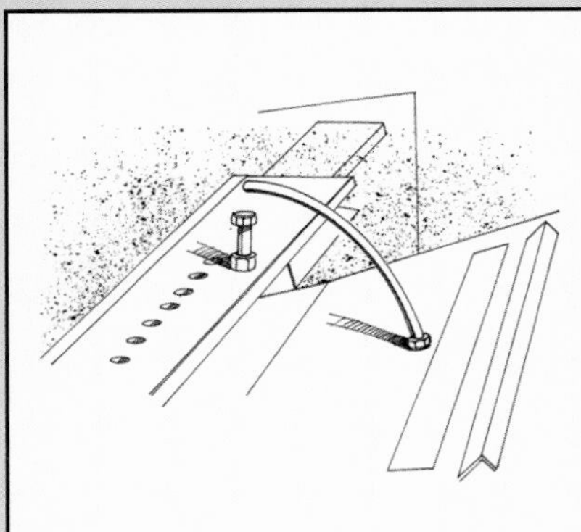

Level indicator

For just $2 in materials, I can tell now if the head is perfectly level with my combine. I attached a 3/8-inch bolt to the tilting head bracket and a 3/8-inch threaded rod to the feeder throat that arches over the bolt. The operator in the cab can detect the slightest tilt by comparing their positions.

–V.L., Iowa

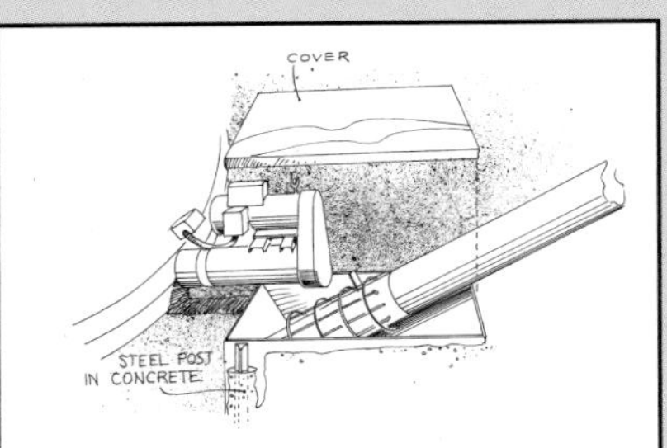

Durable auger sump

Our grain bins have such low foundations we have to dig holes to get the auger below the unloading auger. I built sumps of 3/8-inch steel that won't bend if hit with the auger. They are anchored on each side to fence posts, and the posts are cemented in the ground. No chopping ice in the winter.

–D.W., Minnesota

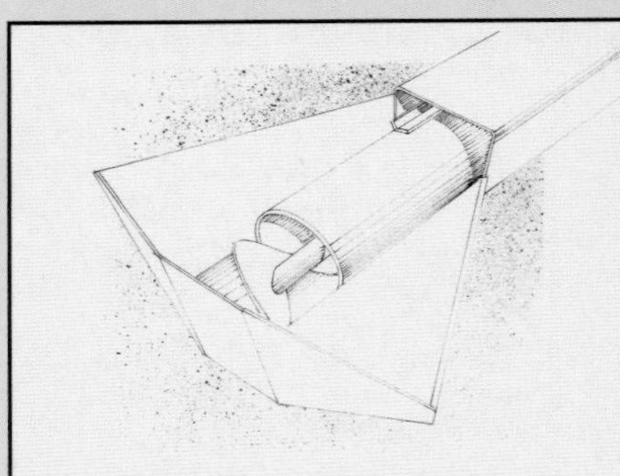

Steadier grain flow

We installed a length of PVC pipe over the intake auger of the feed mill that is 8 to 10 inches shorter than the auger itself, but the same diameter. It's bolted on the sides of the trough with two 5/16-inch carriage bolts.

Instead of the grain rolling over the auger at the bottom and barely dribbling out at the top, we now do our grinding in one third the time.

–T.T., Minnesota

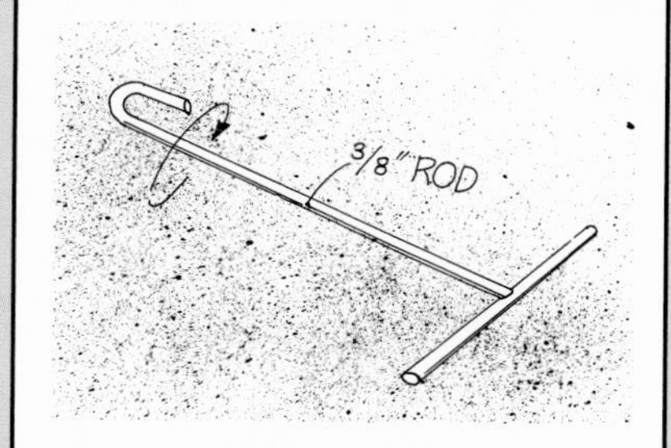

Hay sampling hook

It's always nice to show what's inside the bale when you're in the hay business. Good bales were tough to get into until I made a J-hook out of a ⅜-inch rod. I bent the end and put a T-handle on the other end. Push it in, twist a quarter turn, and pull out a sample of hay.

–G.Y., Minnesota

Stabilize the load

Orange safety fence placed along the length of the load of hay helps keep it from shifting and falling off. We use eight-pack bale clamps. After three rows are in place, we just unroll the fencing down the center, place another row of bales on top, then put one eight-pack on each side of the trailer, shifting half a bale for tying on the top ones.

–J.C., Montana

Low twine-supply indicator

We put on three rolls of twine at a time – two green and one yellow. First we hook up the green twine to the baler, then attach the yellow to the green. When it starts to bind the bales with yellow twine, we know we're running low.

–L.W., South Dakota

Remote control for hay elevator

When stacking in the haymow, I can't always keep up with the elevator speed. Now, instead of having to get the unloader's attention, I shut it down with a simple remote-control relay in my pocket. The switch is at the outlet, and the elevator is plugged into the switch.

–Posted by GuyG

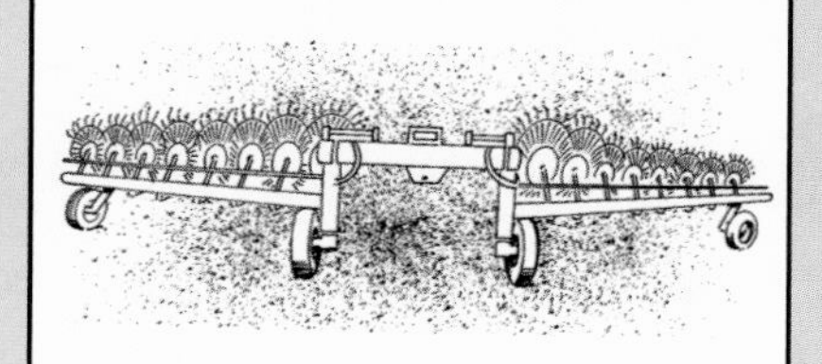

Tandem hay rake

From the arms to the front-pivot wheels, I made everything for this rake in my shop. It is primarily constructed out of two three-point hitches. Folded out, it makes a 36-foot swath, or it can narrow down to 27 feet for heavy hay. When something on it breaks, I build a new one.

–D.B., Missouri

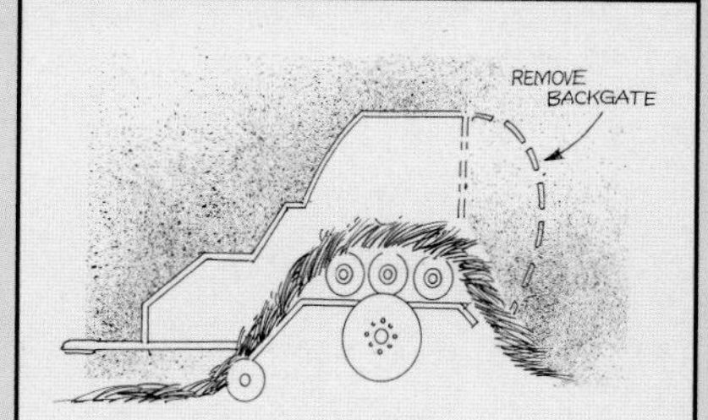

Fluffing up wet hay

Since hay needs air underneath it to dry, we've found an inexpensive way to put up rained-on hay. Remove the back gate from an old roller-type round baler (it takes about an hour), and run over the windrow. It lifts the hay and makes it fluffy so it dries faster. Handy during a rainy haying season.

–M.H., Montana

No more lost pocket knives

I drilled a hole in the wooden handle of a curved linoleum knife and ran a leather strap through it. The strap slips over my wrist. It's convenient and effective at cutting twine off of large hay bales – even if the string is frozen on the bale.

–L.W., Kansas

Easier than manual adjustments

I have modified my conventional tedder by adding a hydraulic cylinder. It makes adjusting the pitch from the tractor a breeze. I can also bring it all the way up to travel pitch when I don't want to throw hay or drag hay on the ends.

–Posted by hay wilson

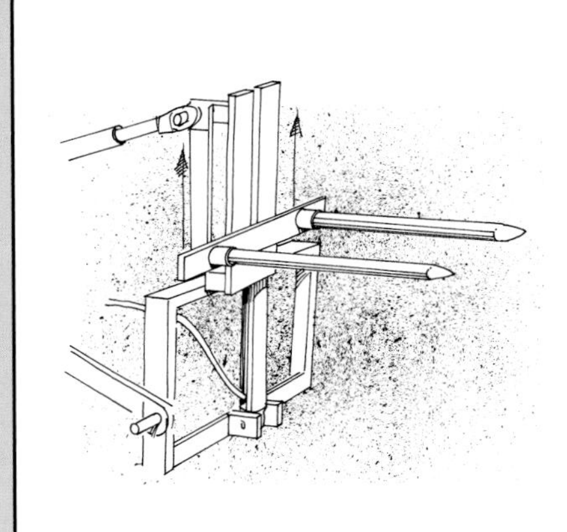

High-lift bale mover

Our regular bale mover would not lift high enough to load wagons. We mounted the spears on a slotted plate, which we attached to a 3-inch hydraulic cylinder with a 16-inch stroke. The cylinder works independently from the three-point hitch.

This additional lifting height allows us to more easily load bales on a wagon in the field with less hassle.

–S.M., Indiana

Preservative applicator

I made a nice and very inexpensive dry hay preservative applicator for my baler. An insecticide unit from a planter is powered by a car's windshield wiper motor. The drive sprocket from the planter fits into the motor shaft. Using metal epoxy helps to prevent slipping.

–K.M, Indiana

Tool makes unclogging a baler much easier

My baler clogs sometimes, and I used to have to turn the flywheel backward by hand to clear it out. A tough job. So I made a 5-foot lever with two small hooks that fit into two adjacent holes in the flywheel. This gives me much greater leverage. I put a holder for it on the side of the baler.

–Posted by neal

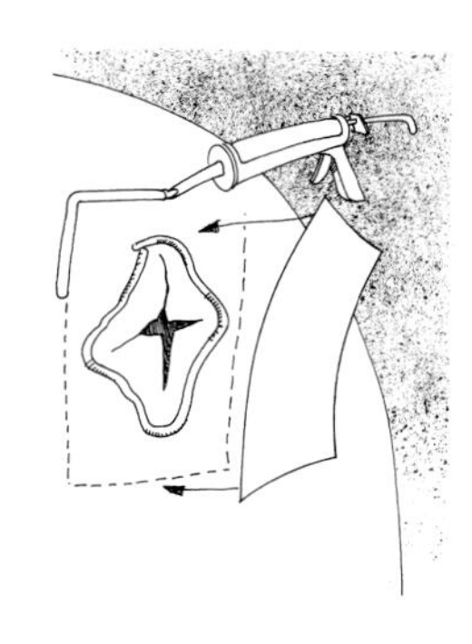

Quick repair for airtight seal

Repairing agfilm (Ag-Bags) with agtape can be frustrating, and it rarely sticks for long. But using silicone caulking is fast and easy. Just fill in small holes (like from bird pecks) and go. For bigger holes and rips, make patches from a piece of used bag. Put a bead of caulk around the hole first, then caulk around the patch for double assurance.

–M.V., Montana

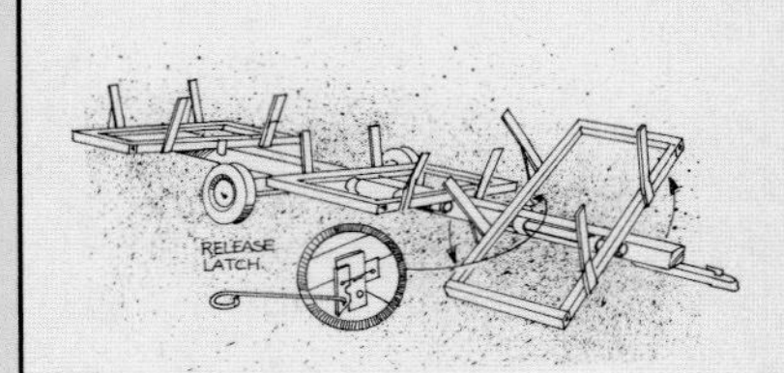

Self-unloading trailer

My welding students built this trailer for a farmer who wanted to carry up to three rolls of hay about 12 miles to feed. He didn't want to leave a tractor or extra hay behind. He also needed to transport it with a pickup. Each cradle dumps separately when the operator trips the latching system.

–W.H., Alabama

Extending life for floater skid plates

The floater skid plates under my haybine were getting worn to the point of cracking. So I added new surface to the skids by welding on leaf spring pieces of the right size. They are wearing well.

–Posted by farmkid

Faster stacking in less space

I set the bottom layer of bales on the end (curved side out) and the second and third layers with the curved side down. I get one more bale per row, and the air can circulate better. Also lets us stack faster.

–Posted by Raleigh

Lawn mower deck cleaner

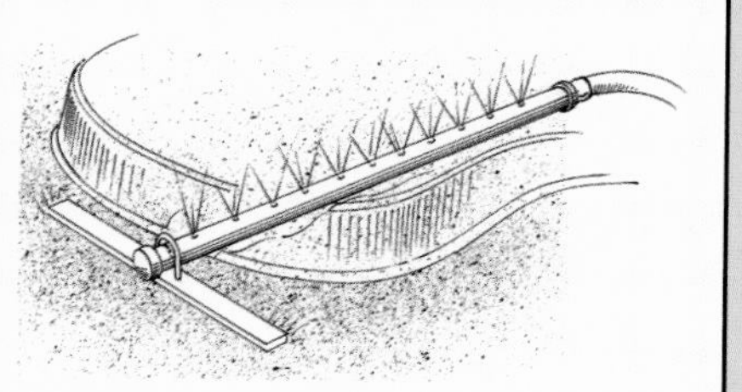

I cut a 1-inch pipe the same length as my mower deck is wide, then drilled four ⅛-inch holes spaced evenly into the top. I capped one end and put a garden hose adapter on the other. The capped end is attached to a 4-inch piece of flatiron with a U-bolt.

I drive the mower over it, moving the deck forward and back to clean the entire undersurface.

–R.K., Nebraska

Water where the roots are

I found an easy and more accessible way to water my melon patch. I attached a 9-foot-long bamboo fishing pole to the outlet end of the hose with duct tape. Conduit would work, too. With 2-foot-high sticks marking every hill, I can water all the vines without stepping on them.

–W.P., Minnesota

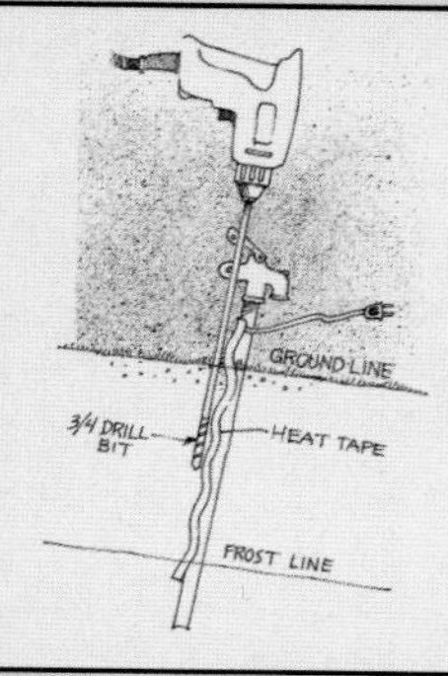

Thawing out a frozen yard water hydrant

Last winter my yard water hydrant froze. I took a ¾-inch steel drill bit and welded it to a ½-inch rod about 36 inches long.

Then I drilled a hole down one side of the hydrant through and past the frost line. I put in the heat tape, and in a few hours the ice was out.

–G.R., North Dakota

Clever use for steel-wheeled farm machinery

We took an old steel wheel from a farm implement about 3 feet in diameter and fashioned it into a unique firewood storage unit for our cabin. It serves the purpose well and is quite a conversation piece.

We welded a 4x4 angle iron brace on the bottom as a base to hold it upright.

–M.M., Idaho

Seed bag makes a lawn tractor cover

I discovered another use for a bean seed bag that I nearly discarded. I found it makes a great water-resistant cover for my lawn tractor. Just cut it down one side first and fit it over the top.

–M.L., Illinois

Easier berry picking

My husband put boards under the legs of a folding seat and added foam padding. With his chest in the seat, his hands are free to pick strawberries. He wears knee pads and pushes the container on ahead of him. He's found this method prevents back strain.

–W.P., Minnesota

Homemade string trimmer

This machine is constructed from a three-wheeled rider mower, a 5-hp. engine from an old sprayer, and the handles of an old garden tiller. After a few hours with a cutting torch and welder, I had a new string trimmer.

I only had to buy two idler pulleys, one belt, and a spool of .155 diamond trimmer line that comes in 24-inch lengths.

–M.P., North Dakota

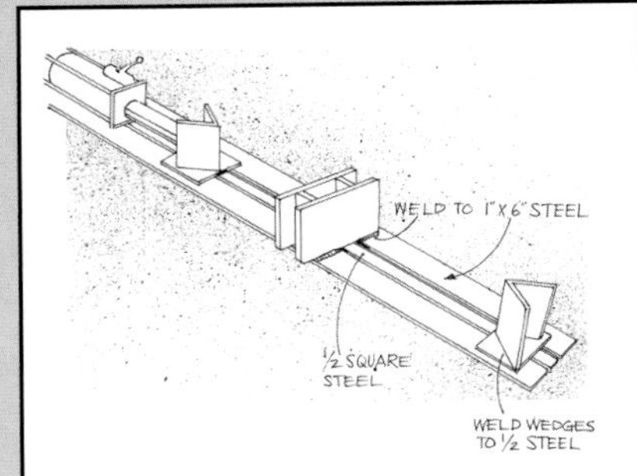

Faster log splitting

This double-stroke log splitter splits wood with each stroke. A 24-inch hydraulic cylinder pushes the guides, which slide together on a 1½x1½-inch square shaft connecting the two wedges. The stop block is made of ½-inch flat steel welded to a 1x6-inch flat steel frame.

–K.S., Wisconsin

Ice mowing for cattail control

Cattails were taking over my pond, so I went out on the ice and cut off the tops so the runoff flooded them out the next spring. This is an environmentally friendly way to control cattail for a number of years since they don't survive well in water over 3 feet deep.

–Posted by tholmquist

Homemade lawn edger

I lay a pointed garden stake, about 6 feet long, over the right side of my lawn mower, extending the point until it touches the sidewalk in front of the mower. The stake lifts the grass and creates straight edges along sidewalks or driveways.

–V.S., Missouri

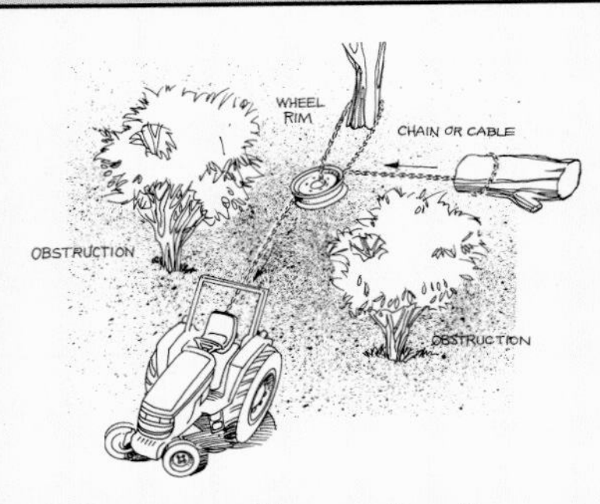

A free pulley block

A steel wheel rim works well as a large pulley to pull fallen trees or other large items from inaccessible spaces. A chain goes through the wheel rim 12 inches from the anchor. Another chain goes around the rim between the load and the tractor. Even hooked chains slide easily around the rim under load.

–J.A., Minnesota

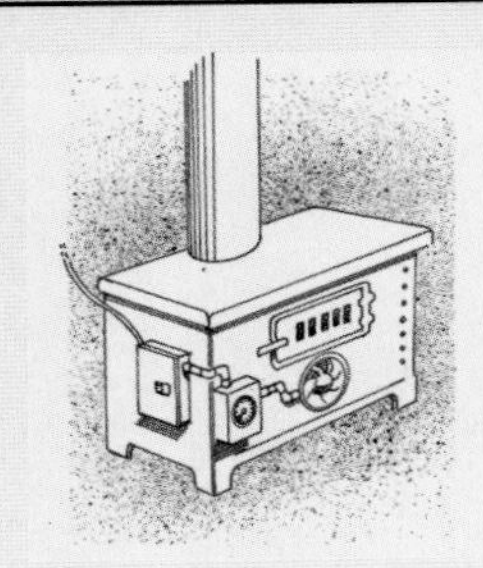

Cleaner chimney

I wired a 110-volt 10-minute on/off timer that once controlled the exhaust fans in a hog barn into the power wires going to the draft blower fan on the woodburning stove. The thermostat is set at 80°F., and the timer regulates how warm the house gets. Now the chimney has less creosote buildup.

–A.H., Nebraska

Fewer nagging doubts about hydrant

To remind myself that I've left a hydrant on, I keep a snap hook with ring and cattle tag attached to the hydrant. Then when I turn it on, I snap it on to my belt loop. It sure saves me from second-guessing.

–T.T., Missouri

Whose boots are whose?

When all our rubber boots were lined up in the barn, I had trouble figuring out which ones were mine. So I took an animal ear tag and pierced it to the outside top. We each got our own color and can even write our names on the tags.

–T.W., South Dakota

Pallets for wood storage

Last year we built these simple firewood racks to save time storing and moving wood. We nailed three pallets together (two 4x4 pallets on the ends and a 4x6 pallet for the bottom) with 2x4s to tie the ends together. We use a forklift on the back of the tractor to move the whole rack at a time. It keeps the wood off the ground when drying.

–S.M., Indiana

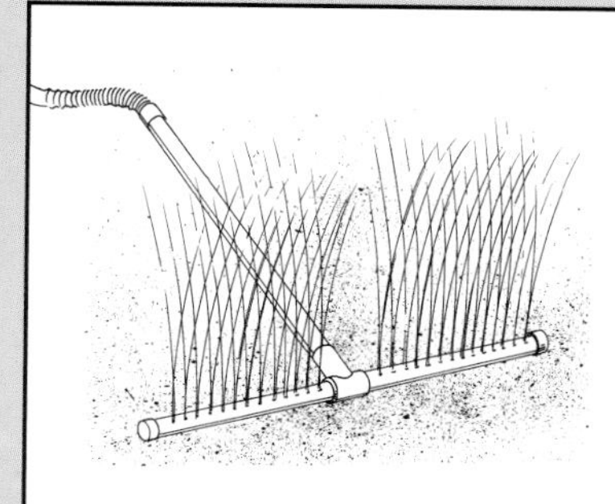

Power lawn sprinkler

Drill ¼-inch holes 2¼ inches apart in 2-inch PVC pipe. Hook the pipe to a 1,200-gallon tank. Start the motor on the tank, and it will shoot water 30 feet or more. Works great for starting a new lawn or in places where you have no other means of getting water.

–H.F., Minnesota

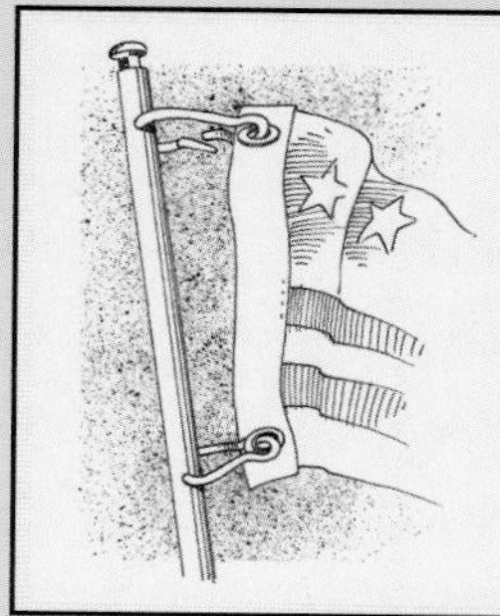

Flag pole built to last

I wasn't happy with the cheap flag poles they sell at the store, so my husband made me one himself.

He used ¾-inch pipe for the pole and finished off the top with a glued on round-headed bolt. The flag is connected with heavy-duty cow clips.

–J.H., Wisconsin

An easy way to get rid of stumps

Loose salt on top of a stump will kill the stump in about one year's time. It will burn down deep into the root area and leave no equipment tracks in the lawn. Works great, and you can burn later.

–Posted by richleon

Tree saver

To reduce seedling mortality from grass competition and mower damage, we surround each tree with a 2-foot- diameter section of waste bale sleeve plastic and hold it in place with a sidewall cut from a used tire.

It mulches the soil, smothers out the grass in a 2-foot circle around the tree, and provides a row marker for mowing.

–B.B., Indiana

Theftproof bait

To keep mice from escaping with the bait, I drilled two tiny holes in a kernel of corn. The holes are threaded with very small flexible wire and fastened to the trap. Since the mouse can't remove the bait, it trips the trap and is caught.

–D.N., Indiana

On three wheels

My wheelbarrow is more useful for me since I welded on the axle from a garden cart about 1 inch lower than the bottom of the skid. It levels it up to hold more of what I put in it. I can negotiate steps and cross gutters more easily, and I can push it hands free.

–E.H., Wisconsin

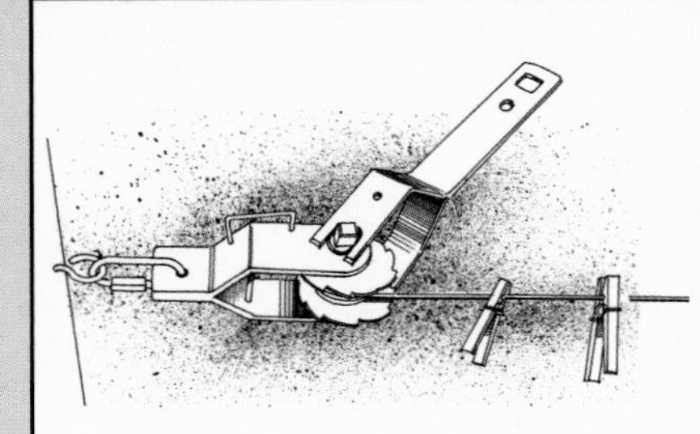

End sagging clotheslines

We added an inline high-tensile fence stretcher at the end of our clotheslines. This keeps them straight and taut. Be sure that your end poles are anchored in concrete.

–D.K., Illinois

Defending the roses

We had a problem with dogs digging in our roses, so I bought 1-pound containers of chili powder and mustard powder and sprinkled both liberally where the dogs had been digging (a 20-foot-diameter rose bed). I've gotten about three good months of no digging.

–Posted by Farmerbr11

Seeing the seeds

I started to plant some alyssum and snapdragons, but the seeds were tiny. So I mixed each one with a ½ cup of sugar and put it all in a salt shaker. The white sugar showed up very well in the black dirt.

–W.L., Iowa

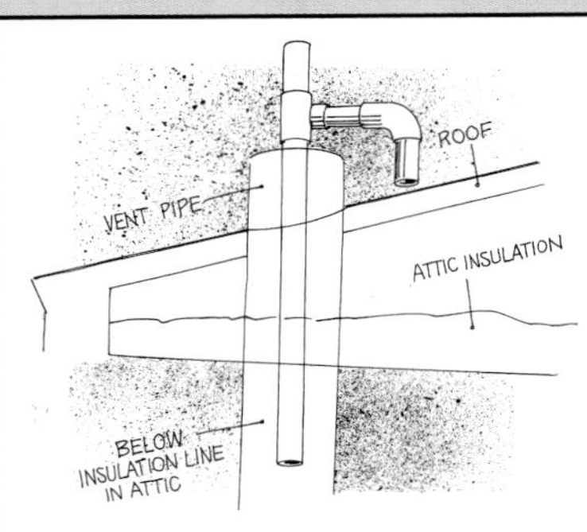

Prevent sewer vent freeze-up

Solder a T-joint 6 inches from the end of ½-inch copper tubing, with the down pipe reaching below your attic's insulation line. Then solder a side pipe and an elbow with 2-inch pipe. Placed in the sewer vent pipe, the copper tubing collects heat and transfers it to the T-joint to vent outside.

–G.L., North Dakota

Firewood holder

Here is a quick and inexpensive way to build a firewood holder. Simply set two T-posts 12 inches apart and another pair as far from the first as you need.

This method holds more stacked wood than any rack I've found on the market and costs less than $15. It holds up year after year, and I can move it wherever I want to store firewood.

–P.S., New York

Cap organizer

We hot-glued clothespins to a 1x4 piece of wood and hung it on the wall to hold caps. We attach the clip to the crown of the cap so the logo is visible and it won't damage the bill.

–R.Y., Missouri

Long-lived work clothes

My duck cloth work jackets and overalls were so faded, I wouldn't wear them to the local feed or hardware stores. But they were still in good shape, so I tried dyeing them. The first night I wore them to my job with the highway department, I fooled everyone with my "new" work clothes.

–Posted by sbradley

Auxiliary garden tool storage

Keep an old mailbox on a post at garden's edge for pliers, pruners, other hand tools, gloves, and even seeds. Saves numerous trips back to the garage or shed. We picked ours up cheap at a rummage sale.

–R.C., Kansas

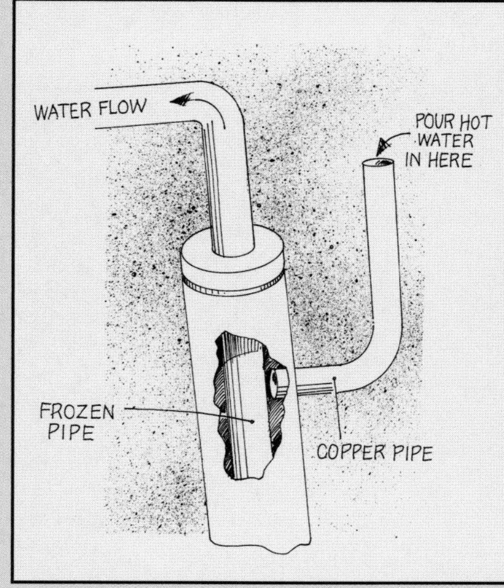

Thawing frozen cattle waterers

I used my torch to warm up a piece of soft copper tubing enough to melt through the outer, larger capped 2-inch plastic pipe over the main water line in the fountain for my cattle.

I left it right in the hole I melted, so now I can pour hot water through the copper tubing to warm the inside line, and water can flow again.

–R.K., Minnesota

Warm calves in cold climates

We slip worn-out sweatshirts cut off at the sleeves onto the front feet of newborn calves that the mama cow refuses to lick. The zipper goes down the back, and we tuck the hood in for extra warmth.

–G.H., New York

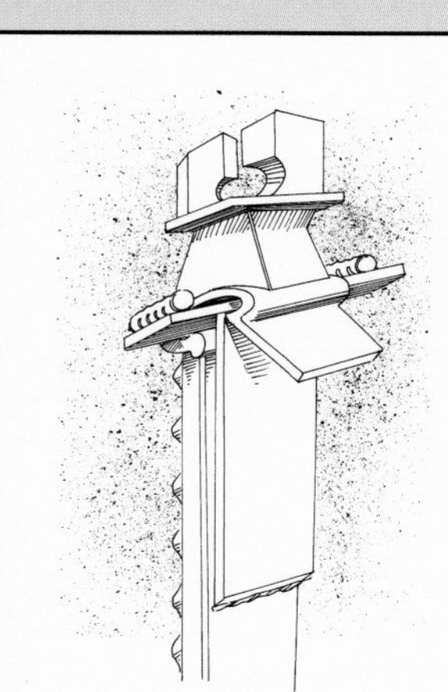

Insulator bracket

These brackets attach electric fence insulators to the back sides of steel T-posts. The 1-inch widths of 3⁄16-inch flat iron vary in length, and 3⁄8-inch rebar keeps the insulators on the brackets.

I weld them onto steel posts in just minutes with a portable DC welder, saving hours of repair to the woven wire that my cattle love to scratch on.

–D.S., Washington

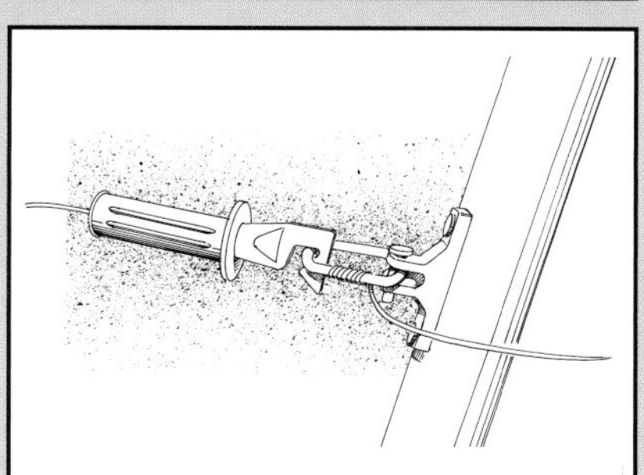

Heavy-duty eye

I couldn't find a company that makes eyes for a heavy-duty electric fence gate. So I made my own using 1⁄4-inch chain quick link and a heavy wood post insulator. A chain quick link opens so it's easy to wrap electric wire around it. Eyes made of wire usually bend out of shape.

–A.J., Nebraska

Keep the clothes on the line

We attached a 5-foot length of 1⁄4-inch chain to the ends of the clothesline. Now clothes go on hangers, which hook through the links. They won't blow on the ground or slide along the line. We enjoy outdoor fresh clothes without the clothespin marks.

–E.O., South Dakota

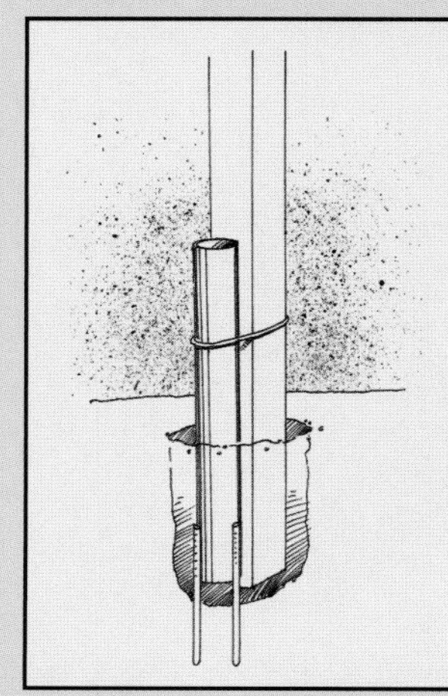

Help for setting posts

To help level and hold fence posts in place, I weld two 10- to 12-inch pieces of 3⁄8-inch rebar to the bottom of each pipe. Now I can set posts by myself without worrying they'll move out of place while I'm pouring the concrete to set them in. I just dig the holes a little deeper than usual.

–K.U., Texas

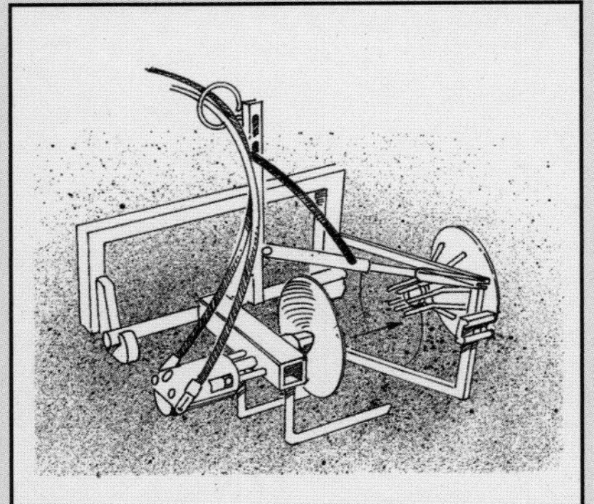

Power fence rolling

We had miles of old wire fence to clean up, so we made a hydraulically powered wire roller. The conventional, PTO-driven wire roller wasn't heavy-duty enough for our purposes.

We also attached a hydraulically powered wire guide to guide the wire from side to side to form nice, even rolls. It's very fast and sure beats doing it by hand.

–J.W., South Dakota

Posthole digger carrier

This carrier makes it easier to mount the digger on the tractor and solves the problem of moving to various locations. Pins lock it in place for transport. A scrap tractor PTO shaft welded to the frame keeps the PTO shaft in place during transport. Square tubing uprights serve as storage.

–R.K., Missouri

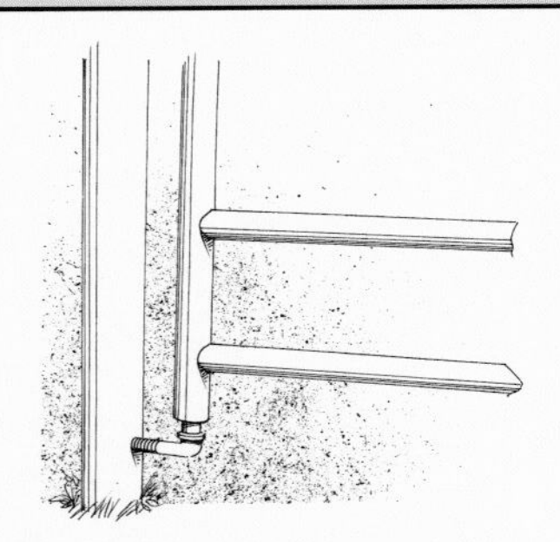

Boost for sagging gate

Livestock gates sag after a while. We screwed an extra gate hinge lag screw to the bottom of the gate. The gate now rests on the lag screw. We have a post set with this rest whether the gate is open or shut. It's an improvement over a wooden block and provides support if a calf tries to push on the gate.

–K.W., Illinois

Calf trailer

This calf trailer fits behind an ATV. It is 4 feet square and utilizes a 16-foot cattle panel. The cradle inside holds a calf for tagging. Baler belting protects newborns.

–G.M., Missouri

High visibility for bale feeders

Feeding in the dark is a lot easier since I put reflecting paint on the top bars of my round and square bale feeders. The loader lights illuminate them as soon as I drive into the yard. I painted the corner posts and gate areas on my corrals, too.

–D.H., North Dakota

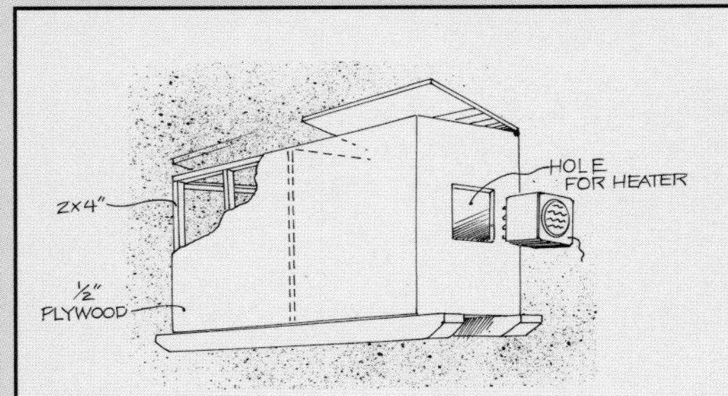

Warming box for calves

My hot box is made of ½-inch plywood and is 8x2x4 feet with a top on it. I put skids under it for ease of movement. It is divided in the center and will hold two calves. I put small blower heaters in either end of the box. Several neighbors have brought chilled calves to me.

–R.P., Oklahoma

Night vision ID is a big help at calving time

We use reflective tape on the cow's ear tag for the numbers. It makes it much easier to identify a cow at night.

The reflective tape shows up real well in the dark when a spotlight or a flashlight is shinning on it. It has worked very well for us especially during calving time when we are constantly checking cows at all hours of the night.

–M.M., Iowa

Switches in easier reach

We've run string from one end of our milking parlor to the microswitches that control the electric motors on the other end that open and close the parlor's three doors. Now we can stand anywhere in there and open or shut the doors by pulling on the string.

–N.G., Ohio

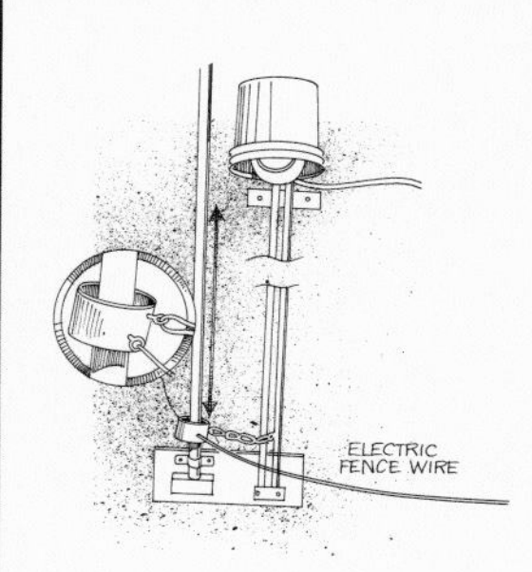

Automatic lift for fence

I don't have to jump off and on the skid loader to get in the cattle yard since I mounted a garage door opener on the silo to pull the wire up and down. The loader's key chain has a remote opener. I mounted 1½-inch pipe on the silo with a 4-inch PVC collar that slides up and down as the opener moves.

–J.B., Iowa

To eliminate animal odors

To eliminate strong dairy farm (cattle and manure) odors from hands, scrub down with toothpaste. Of course, gloves and a good barrier cream work well as preventive measures.

–Posted by cowvet

Windbreak without swirling wind or snowbanks

I'm making a windbreak with some 9-inch irrigation pipe we don't use any longer. The cracks seem to let through just enough air and the cattle love it. There's no swirling wind or snowbanks. I use Forever Post spaced every 6 to 8 feet, buried 4 feet in the ground on the outside of the fence.

–Posted by don

Cows bond with syrup

When a cow loses her own calf, it's hard to get her to bond with another. We put some pancake syrup on her face, then on the calf's face, back, and rear end. She'll lick the calf clean because it tastes sweet.

–B.L., Iowa

Portable calf ring

To make it easier and safer to catch and work with newborn or young calves when they're out in the field, we came up with this idea. Wrap a cattle wire panel around an old hay feeder ring and wire it to the feeder. When you need to catch a calf, use the tractor loader to pick up and carry the ring, place it over the calf and drop it around the animal. This is an easy way to contain the calf, and it serves the keeps the cows away while we're working on the calf.

–S.Y., Missouri

Calf walker

If you have to transport a newborn calf a short distance, it's much easier to let it walk instead of having to carry or haul it.

A 6-foot piece of ¾-inch heater hose looped around the calf enables you to turn it to the right or left. And there are no objections from cows with strong maternal instincts.

–R.B., Missouri

U-bolts anchored in tie stalls

When a new heifer comes in to milk, we halter and tie her tight to one of the large, heavy U-bolts cemented against the wall in our tie stall barn. About 99% of new heifers learn to milk in three days with no kicking at all. This also works well for vet work.

–J.W., Wisconsin

Large-capacity mineral feeder

Use 55-gallon plastic barrels mounted in truck tires with stainless-steel bolts. I attach eyebolts to the tires so I can pull them around with my four-wheeler. The tires keep them from turning over.

–Posted by Norman

Hog chute to calf toter

We took an old hog catch chute and turned it into a portable calf cart for catching or hauling newborns from the pasture to the barn.

We replaced one side with cattle panel so the cow could see and follow the calf in the cart. We added a tongue and two wheelbarrow wheels to make it portable behind our ATV, tractor or pickup. The catch chute is handy for vaccinations and tagging. We have less than $100 invested.

–B.P., Iowa

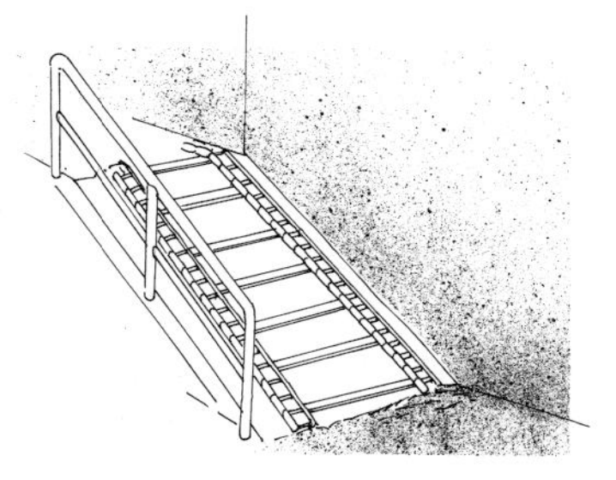

Sure footing on chute

We salvaged the live floor from an old manure spreader and bolted it over our slippery loading chute. To attach the live floor we drilled holes in the concrete ramp and live floor web, making sure to mount the chain upside down. The only expense in the project was for 24 concrete lag bolts.

–J.M., Pennsylvania

Corner posts that will outlive you

We use plow beams off four- or five-bottom plows as anchors. You can buy a five-bottom plow for $50. We cut the old iron off, dig a hole 60 inches deep with an 18-inch bit, then cement the beam in. No brace post is needed. We did one 22 years ago, and it still looks the same today.

– Posted by garvo

How do you get rid of pesky fly problems?

I ordered fly parasites from kunafin.com (800/832-1113). I actually started late in June and thought it was too late, but noticed a huge difference in just four days. I spend $20 every two weeks for 15,000 parasites. You now wouldn't know there is livestock around the place.

–Posted by Donna

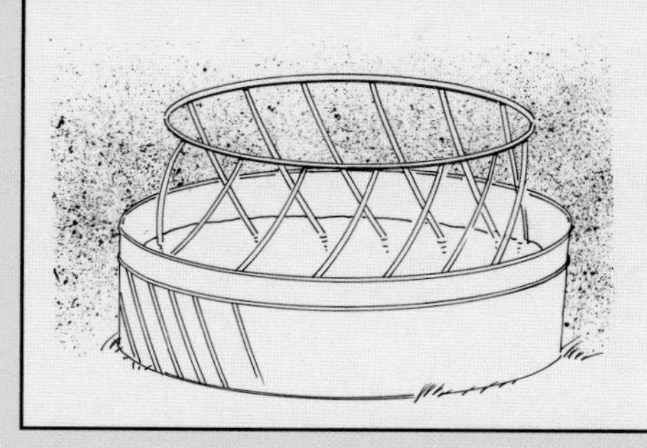

No more cows in water tank

When we were hauling water to temporary galvanized water tanks, cows were always crowding around, getting into or destroying the metal tanks.

We found that placing a bale rack inside the tank keeps cows and calves out of the tank. The ring needs to be sized to the tank.

–R.S., Florida

One-person operation

We needed help catching heifers and calves at vaccination time. This can be a tricky job if there's only one person around to do it, which is often the case. So we constructed this gate that hinges on one end. It swings out from the wall to catch the animal. The gate has worked well for us.

–R.D., Wisconsin

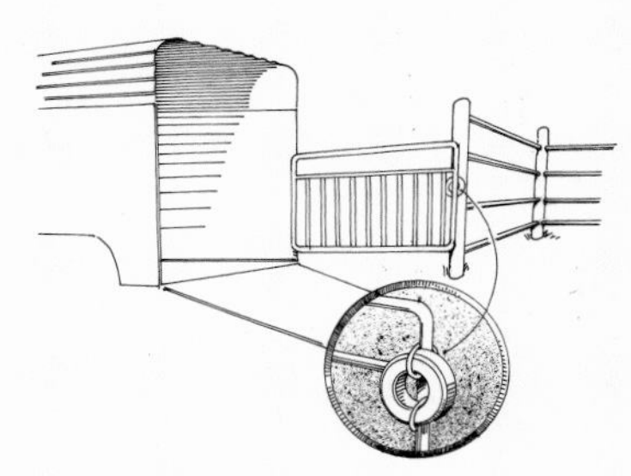

Handy gatekeeper

If you have ever loaded cattle by yourself, you will appreciate this idea. I used a magnet from an old radio speaker on the outside of the tailgate. The magnet grabs the metal fence post and holds the gate securely when we're loading.

Then when I need to close the gate quickly, I can. No more trying to untie a rope or wire while frantically guarding the cattle or other stock from exiting.

–K.U., Texas

Year-round drinking tub

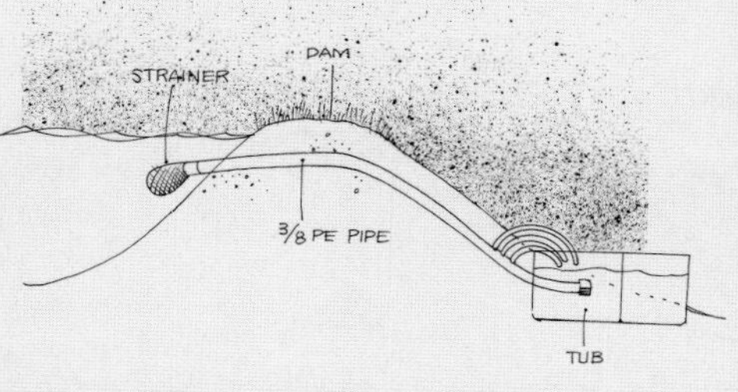

This system doesn't freeze because the ⅜-inch polyethylene pipe is buried, and the tub is set in-ground. Four short drip tubes are wrapped in an insulating sleeve, and at least one tube is always running. The .06 drip tubes are plugged shut with thumbtacks and unplugged according to need.

–G.B., Tennessee

Drive-through ATV gate

My grandchildren forget to close the pasture gate when they're riding ATVs. So I cut heavy indoor/outdoor carpeting into 10-inch-wide vertical slats, folded them over, and riveted them onto a piece of pipe attached to a couple extra-tall fence posts. The cows won't fool with it even if the wind blows it around.

–T.T., Kentucky

Combination cattle oiler and mineral feeder

This unit is made from an old four-wheel running gear with a 2x6-foot wooden box fastened to it. I welded on steel studs and attached a corrugated steel roof to keep the salt and minerals dry and the insecticide from washing out. Its mobility works great for my rotational grazing system.

–D.B., South Dakota

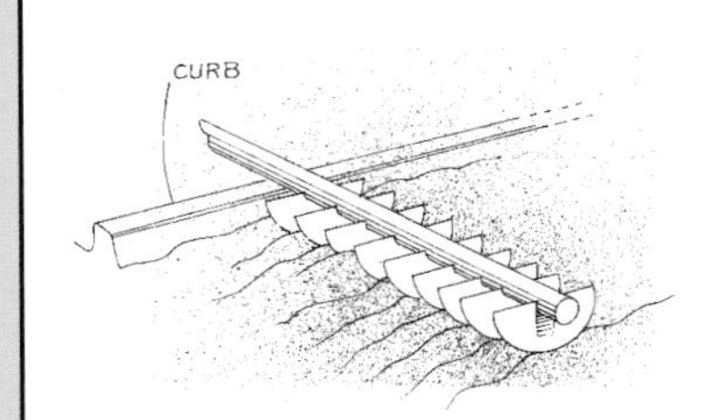

Grooming the barn

Our cows work the sand bedding in the free stall barn from the front to the rear and out over the retaining curb. So we replaced the tines on our skid steer's grooming attachment with seven 20-inch discarded light disk blades.

The blades are cut in half and welded 8 inches apart at a 65° angle. This keeps the stalls in shape and saves a lot of sand.

–F.K., California

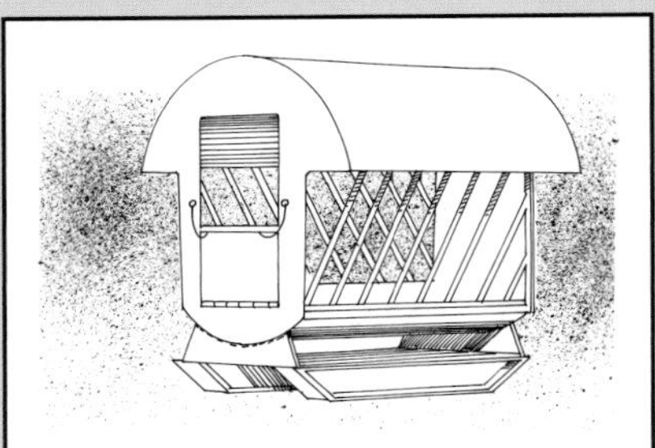

Indestructible hay feeder

Having built several wooden hay feeders that failed the test of time, I wanted one that would last.

I purchased a used, large steel tank. After cutting out the sides and welding in slanted bars above the bunk and a set of runners, I had a sturdy feeder.

The rounded roof has a nice overhang and I cut an opening in one end into which I can put the hay.

–E.M., New Hampshire

Milk cart conversion

I built a plywood flatbed for my golf cart, but it could also fit an ATV. A polydome tank with a built-in motor and attached mixing rod sits in a plywood box, which is easily removed. The tank will feed up to 90 calves with one fill. I also have a gasoline-type nozzle for easy milk feeding.

–S.M., Wisconsin

Long-handled curry combs

Our 4-H calves are pretty jumpy at first, so we took old broom handles and duct-taped them to the handles of our curry combs. This puts us a good distance from the calf. Also works well on their undersides, where it's dangerous to reach anyway.

–C.C. (age 12), Illinois

How do I deal with docked tails on dairy cows?

Our vet recommends it in cool weather, not fly season. He suggests waiting until a heifer has her first calf. Dock tail about a hand's width below the vulva with a tight elastic band and lop off with shears a week later. We use a shot of Naxcel to avoid infection.

–Posted by shari

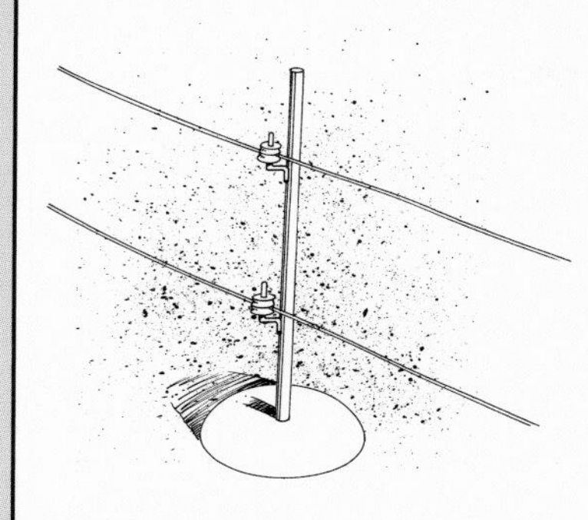

Portable fence post

These 32-inch disk blades turned upside down with 1-inch pipe welded to them make good movable electric fence posts. I use ⅜-inch rod on the pipe for insulators. The corner posts have round, white insulators and there are adjustable insulators on the line fence posts.

–D.O., North Dakota

Alternative bedding material

We've started rowing bean stubble and chopping it for bedding. It lasts longer in the stall than straw and soaks up more moisture. We stack it against a wall with a silo blower, which has a long, curved spout similar to a forage harvester.

–J.B., Illinois

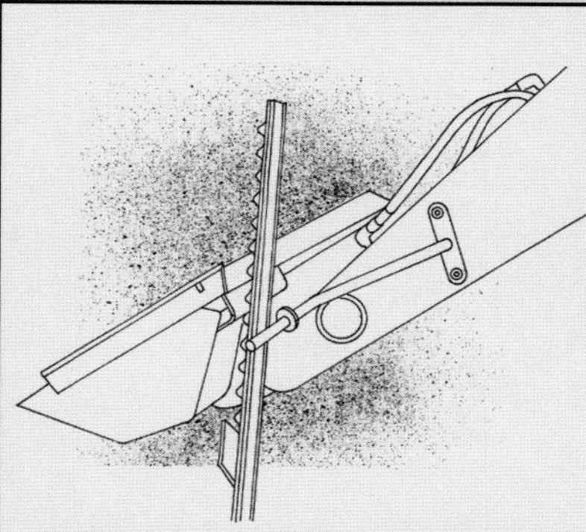

Hydraulic post puller

We mounted an additional L-shape bar to the arm of the front-end loader on our tractor for pulling posts.

You can quickly and easily remove any post by extending the bucket and dropping it over the post between the L-shape piece and the bucket. Curl the bucket down against the post and apply pressure. That subsequently raises the arms so you can extract the post.

–J.A., Iowa

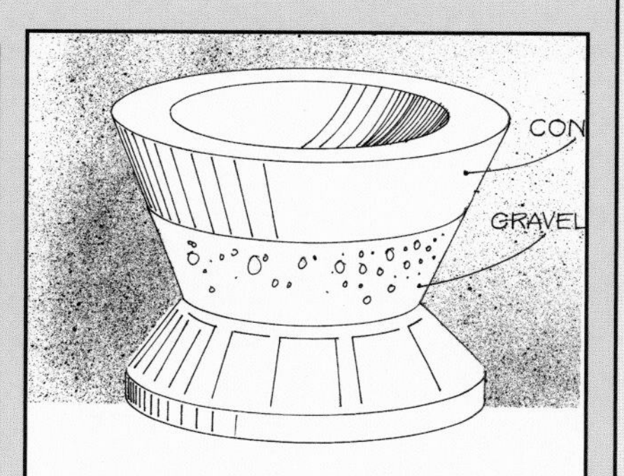

Salvaged hog equipment

We used the bottom portion of an old hog feeder to make a feeder for cattle. We filled the bottom of the feeder with rocks and gravel. We then poured 3 to 6 inches of concrete on top of the rock, (pouring it fairly dry) and hollowed out a bowl in which to feed cattle.

The new feeder is very sturdy and feed wastage is limited. We used nearly ⅓ yard of concrete.

–K.S., Missouri

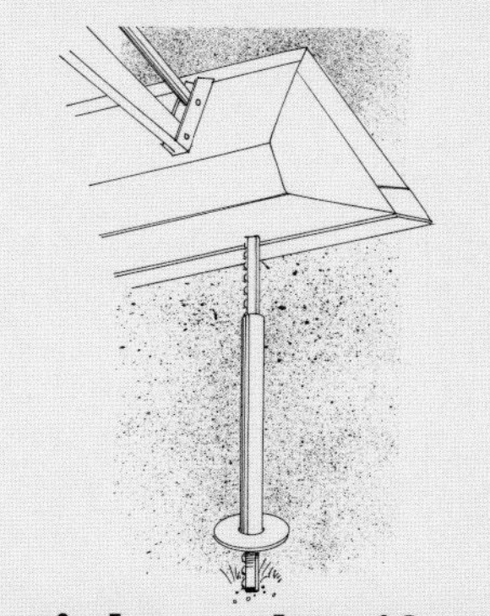

Straight and uniform steel fence posts

I use this device when driving T-posts with my front-end loader. I cut a piece of 2½-inch pipe to the height that I want the posts. I cut a piece of ¼-inch steel about 12 inches in diameter with a 2½-inch hole in the center. This I welded to the bottom to form a base.

I start the post in the ground, then use the loader to push it in the rest of the way so all the posts are the same height.

–J.G., Texas.

Are medications plugging your nipple waterers?

Adding citric acid to the medications solution has worked for some. We find running a diluted bleach solution through the lines helps to break up the dextrose and sucrose that seems to plug the lines. It has helped us immensely and saved a lot of time.

–Posted by Steve

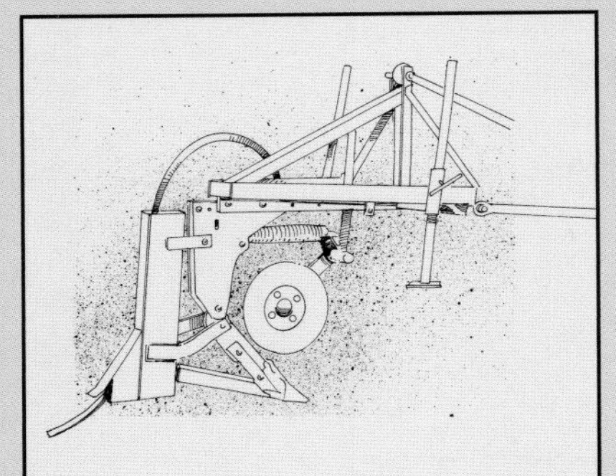

Homemade trencher

I made this trencher from scrap iron. The main frame is an old plow beam, and the point is a shovel. It goes 12 to 18 inches deep and fits on a three-point hitch. The rolling coulter is for cutting sod. Plastic water and electric lines go through 3x4-inch tubing. I have used it in my yard.

–S.R., Minnesota

Mower modification

I took two big swivel wheels and bolted them on the sides of my rotary mower, near the rear. This keeps the sides from dragging the ground when I'm cutting. It helps keep it in good condition.

–E.H., Missouri

Polyshank savers for drill arms

The harrow lift was wearing on our JD 750 drill's closing wheel arms. So we mounted polyshank savers on the arms. Two U-clamps hold each one in place. After three years of use, they show less wear than the steel did in one year.

–M.E., Illinois

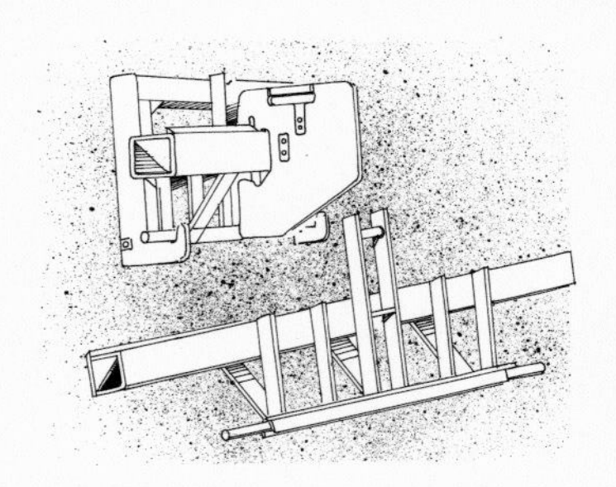

Adjustable weight

To make our tractor as heavy as possible without adding fluid to the tires, we built a frame to attach to the three-point hitch quick coupler that will hold up to 4,000 pounds. It raises the weights high enough to hook up implements without removing the weights.

–J.S., Washington

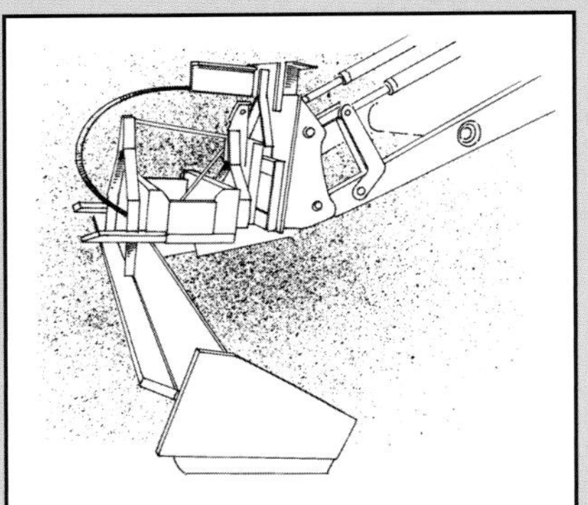

Box blade on loader

Turning around in the tractor seat to operate the box blade was breaking my neck. So I used the front forks on my loader to mount the blade, which sure makes dragging dirt a whole lot easier. I just use a bale clamp to hold the box blade in place, but a chain binder will work, too.

–T.W., North Carolina

Center-pivot track closer

We took the manual-fold wings from a used tandem disk and welded a frame and three-point hitch to hold the wings. The amount of dirt it throws is regulated by adjusting the length of the three-point hitch's top link. The total cost was around $350 after selling the center part of the disk.

–E.H., Nebraska

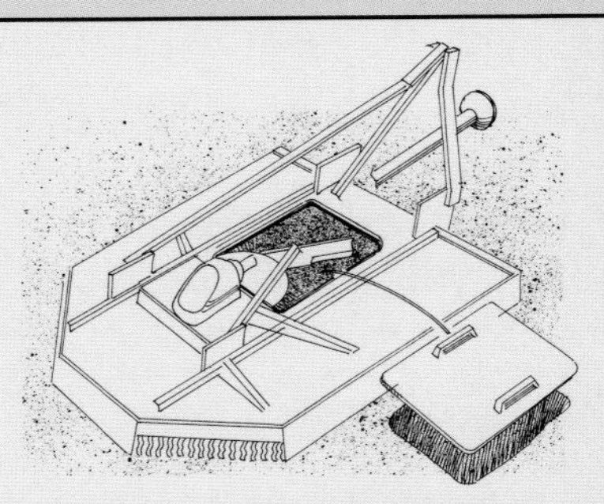

New mower deck lid

There was a badly rusted area on my mower, so I torched it out along the frame and made a lid to slip in instead.

The happy accident was that now there is much easier access to the blades. Nobody minds the task of sharpening the blades now because we can get at them from above.

The cover is made from previously used 3⁄16-inch steel plate.

–W.Y., Kansas

Light to plant by

I mounted two lights behind the fertilizer tanks on my planter and wired them through the flasher plug-in. We needed something because the fertilizer tanks block the tractor lights from shining on the planted ground. These also come in handy when I'm filling boxes in the dark.

–T.D., Minnesota

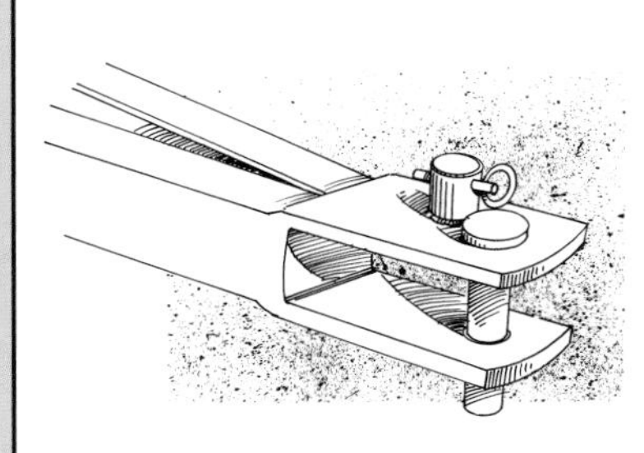

Shorter hitch pin

Drill a hole in a short piece of 7/8- or 1-inch shaft. Weld to the tongue of an implement so when a linchpin is inserted in the hole it extends over the head of the hitch pin. This allows a short hitch pin; it doesn't extend down far enough to drag up a wad of hay.

–P.D., Missouri

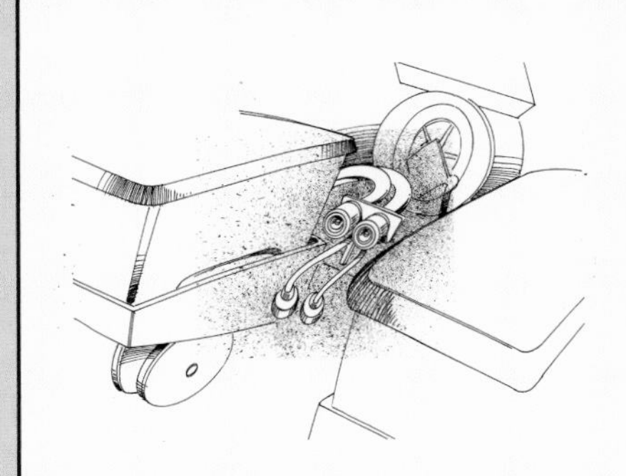

Put to good use

When I started using minibulks for planting, I realized I had an extra hydraulic outlet on my JD 8300 that I didn't use. I ran hoses from that extra outlet to the back of the 7100 Maxemerge planter and mounted them to one of the vacuum impeller mounting bolts.

–T.B., Nebraska

Easy-to-find connector links

I painted all the connector links on my combine's roller chains with a fluorescent orange-color paint. Now when I need to remove a chain, I can immediately spot the connector link. Saves a lot of time scratching on a greasy chain.

–Posted by L. B.

Acrylic plastic enclosure for a skid steer loader

This enclosure for my skid steer loader is nice for pushing snow in cold weather. I used cable ties to hold acrylic plastic on the sides and back. Then I pop-riveted it to the door. The door frame is made from ½x2-inch metal. Four 2-inch hinges hold the door on, and a 12-inch rubber tiedown keeps the door shut from the inside.

–D.P., Indiana

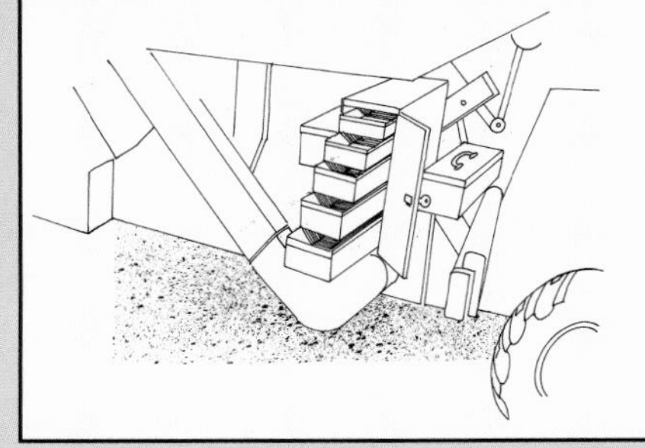

Five-drawer cabinet

I put this five-drawer cabinet on my combine with two more toolboxes, one on either side. The back side accommodates oil and spray bottles for lubricating chains. It also carries hammers, locking pliers, full sets of metric and standard wrench and socket sets, and lots of extra parts.

–R.V., Iowa

Dual-wheel swather

An extra wheel helps my machinery keep from digging in. This is particularly useful when the hay or other feed is ready to come out of the field, but the ground is still muddy or soft. To add the extra wheel, I welded a spindle to the inside of the outer spindle.

–T.T., Kansas

Better wheels for hay rake

The wheels on our new two-rotor hay tedder were too small. So I cut the hubs off the rear axle of a front-wheel-drive car. The frame is made of pipe, installed over the axle, and hubs are welded to the pipe. With spare car tires, it maintains the same ground clearance. Trails great.

–G.K., Pennsylvania

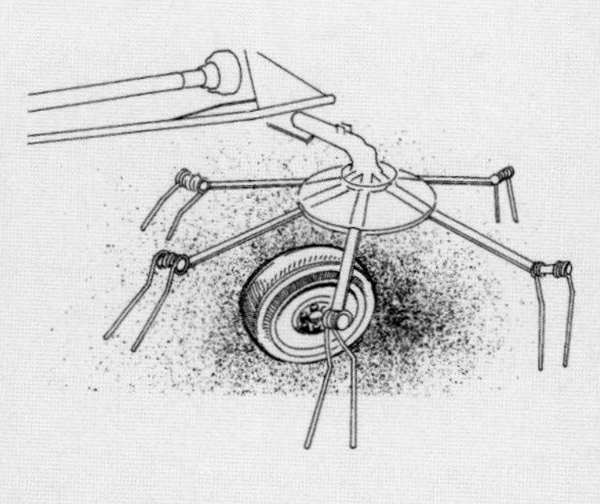

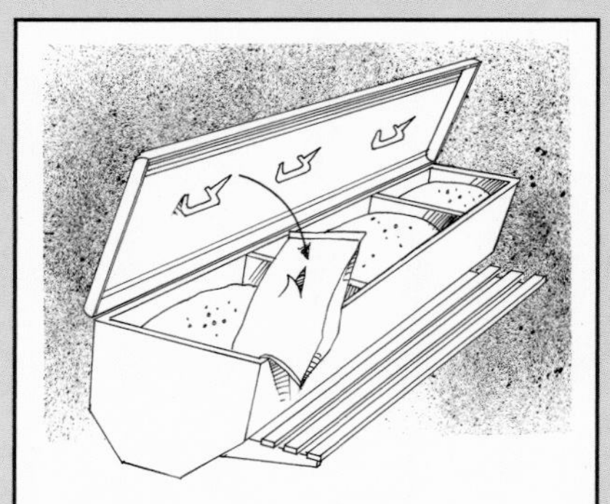

Quick-and-easy bag opener

I welded old silo unloader links to the lid of our grain drill. By simply dropping the bags on the sharp edges, the bags are cut, opened up, and the contents are emptied into the drill. Any sharp piece of metal will do. This saves me a lot of time.

–C.S., Wisconsin

Hands-free pump controls

To operate my pump controls' electric switches more easily, I installed a dimmer switch on the tractor floor. It's attached to 3-inch, flat 10-gauge metal with two small bolts, and it's wired to the battery.

–K.U., Texas

String catcher keeps baler twine off axle

We U-bolted short pieces of ¾-inch round bar with hooks at the end to each axle on our wheel loader parallel to and 4 inches from the axle. This prevents most strings from even beginning to coil.

–J.H., South Dakota

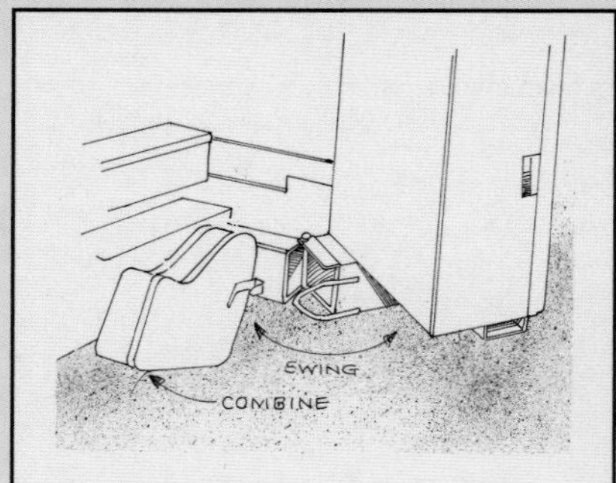

Swing-away toolbox

I mounted my toolbox on a swinging arm near the tailing elevator so I can get in closer to work on the combine. I fastened 4x3-inch tubing to the frame and welded on a hinge so it can swivel. It's held in place with a lock from an old combine tailing elevator lid.

–D.W., South Dakota

Flashing signal light

I added this flashing light to my combine so it is visible to the person driving the grain cart next to the combine at night. It's wired to a switch that I can flip on when the grain cart is full (or when I'm through emptying the combine), so the driver knows it's time to pull away.

–R.V., Iowa

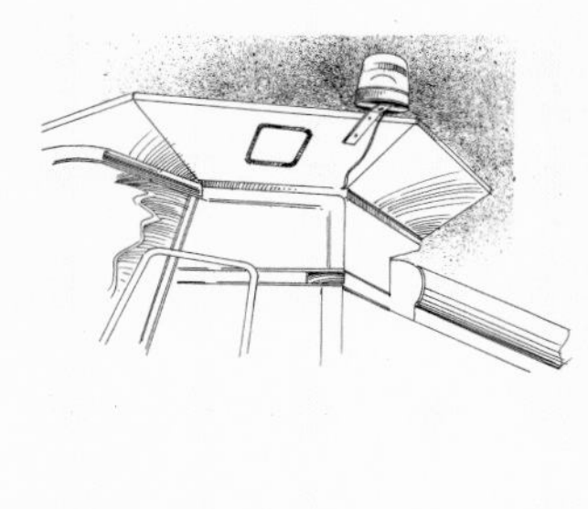

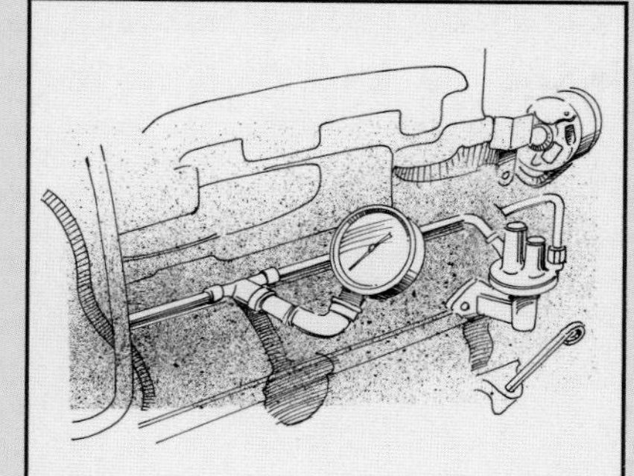

In-line pressure gauge

I placed a tee in the fuel line on some of our tractors. Then I installed a permanent fuel pressure gauge between the prefilter and the transfer pump.

The gauge showed we had two bad transfer pumps. It also shows clogged prefilters, preventing real problems. A 15-psi gauge is sufficient for distributor fuel pumps; for an inline pump, it must read over 20 psi.

–S.J., Tennessee

Electric cylinder

We used to have to climb into the grain tank and manually lower our combine's in-bin auger so it would fit under the machine shed's doorways.

Then we placed an electric cylinder (like one that raises and lowers combine platform steps) under the auger. Now a flip of a switch in the cab gives us the right height for either combining or storage.

–A.B., Ohio

Save trips on and off the tractor seat

I designed a long hitch pin (¾x7-inch pin attached to a length of ½-inch rod) so I can unhook wagons right from the tractor seat.

There is a hook just above the pin to lift the wagon tongue.

–D.M., Iowa

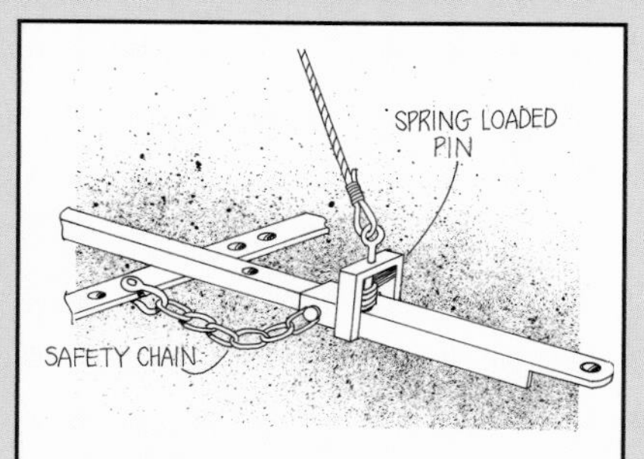

Drawbar extension

This sleeve fits over our drawbar and is equipped with a spring-loaded pin and safety chain. This allows added flexibility when hooking up wagons with heavy tongues. The extension can be hooked up, and it snaps in place when you back up

–W.K., New Jersey

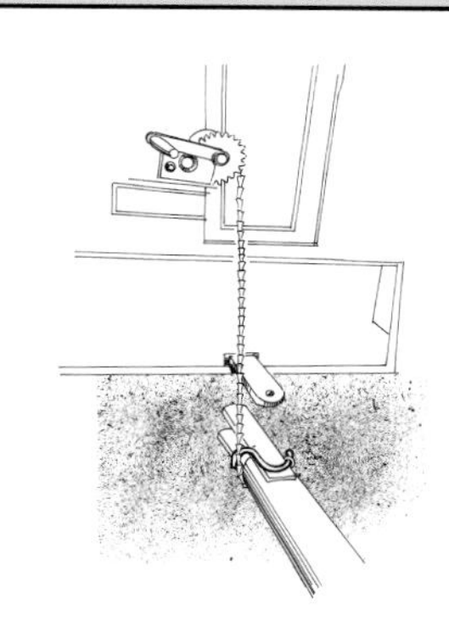

Easier wagon hitching

This arrangement helps me easily hitch my wagons together. The light-duty winch is mounted onto the truck's tailgate, and there it will hold the tongue at the proper height.

Since this system makes the hitch fully adjustable, hookups are now nearly effortless.

–C.C., Georgia

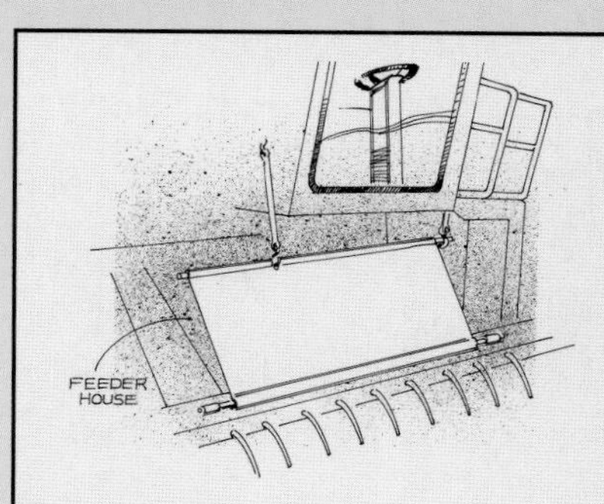

No more trash buildup

We may have our bean head off several times a day, cleaning the trash buildup. So we added a tarp a little wider than the feeder housing, with steel rods running crosswise secured by tarp straps. This deflects beans and dust back into the header so well that we're left with very little cleanup at the end of the day.

–G.M., Illinois

A double rake hitch

An old pull-type sprayer cart serves as an excellent cart for building a twin rake hitch. I started with the basic cart that had a square beam. I then inserted a slightly smaller beam and hitch into the larger beams. They can be adjusted in or out to allow the following rakes to put the hay into the appropriate- size swath. It handles a left and right delivery rake and took longer to paint than to build.

–J.D., Illinois

Twice the tubes on a round baler

I made replacement tubes by welding two pieces of ¾-inch pipe together on one end; the other end is bent 4 inches apart. Then I added extra twine guides for the two strings. The cutter works the same.

–T.C., Oklahoma

Free-flowing wheat seed

The seed hoses on our grain drill kept kinking in the lowered position. A simple wire field flag bent at a 90° angle inserted into the seed tubes and down through the seed hoses lets the seed flow freely.

–G.C., Ohio

Feed bunk scraper

Cleaning old feed and snow out of feed bunks is easier with this handy bucket I made to fit the skid steer. It has swivel arms to hold the bucket. The arms have shear pins that break instead of bend the frame. I put skid shoes on the front and sides of the bucket to align it with the next feed bunk.

–D.B., Wisconsin

Remote auger control

We haul to the bin with an auger cart. We can start our auger tractor with a remote control operated from the seat of the auger cart tractor. When the stationary wagon runs empty, a mercury switch on the paddle shuts the tractor off. It is on a delay timer to let the auger clean out before the tractor shuts off.

–T.B., Illinois

Help for planter modifications

Mounting heavy trash whippers to my planter is a faster, simpler job now. That is, since I welded a stand-up column to my $15 two-tongue jack.

The jack allows me to maneuver the trash whippers sideways and, of course, gives me a wide range of variable vertical settings.

–L.W., Iowa

Heavier scrape blade

My handheld scraper blade just wasn't heavy enough to cut through frozen snow and manure. So I welded a 1-inch pole to the blade housing and put on some flat, stacking barbell weights to make it heavier.

–G.P., Virginia

Inexpensive platform

We store our corn head on a wooden platform made from three old light poles with 2x8s across them. It holds our 16½-foot corn head and works great for moving the corn head in and out of the shed.

–Posted by cornboy9

Homemade snow wing

I used an old grader blade for this project, but a person could fabricate one, as well. One end attaches to the drawbar, and the other end has a cable hooked to the loader bucket by a chain, which makes the blade angle adjustable. Raise the bucket to make a sharp right turn.

–M.B., South Dakota

Easy does it

One of the levers in our two-spool hydraulic valve is blocked off so it can only move one way. Plumbed into the down hose with a gate valve in-line, it's adjusted so it can only lower the load slowly. The other lever moves the loader up and down at a normal speed.

–R.N., Wisconsin

Screens keep chopper clean

Transparent bug screens made for truck grills help keep our self-propelled chopper clean during hay season. I made an angle-iron frame and positioned it horizontally over the windrower. The snaps make it easy to take off, and it doesn't crack.

–T.W., South Dakota

Two-bottom plow from a six-bottom

Since six-bottom plows are getting to be a thing of the past, there are plenty around. I cut the front four bottoms off, and now I have a steerable two-bottom plow for my tractor.

–L.C., Iowa

Portable auxiliary lights

This lightbar (made of lightweight steel, metal conduit, two taillights and clearance lights) has a 35-foot, four-conductor cord with a six-pole plug-in. A volt meter and switches are installed in an old planter monitor box mounted to the tractor fender: push/pull for taillights, momentarily on for brakes and an in-line flasher for emergencies. It can also power a floodlight.

–H.J., Illinois

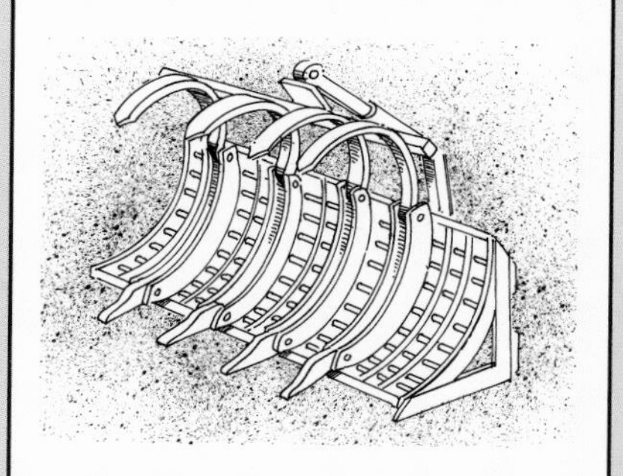

Grapple with this

This tine fork is heavy duty enough to handle rock ripping, concrete, loose manure and moving large round bales. The teeth are chisel points from a digger and are held together by 2x4 box steel beam. The main frame is made of 2x2 tube steel from an old field cultivator. A two-way cylinder opens the grapple fork, and a quick-attach mount plate lets it disconnect easily from the skid steer.

–T.B., Nebraska

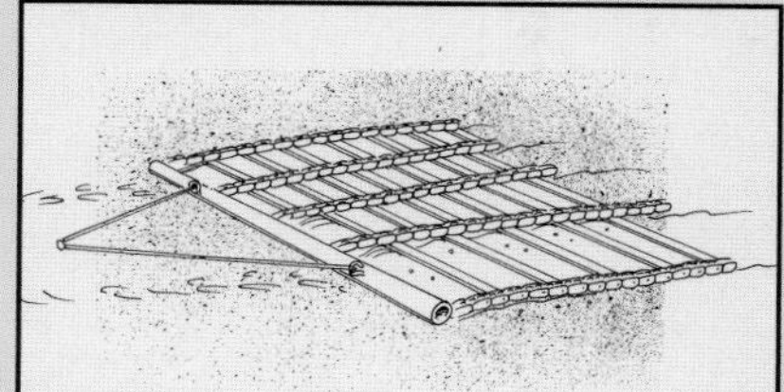

Gravel drag

A halved combine feeder house chain keeps my gravel farmyard groomed. The two sections are connected by a 2-inch pipe that also acts as the leveler. I used ¼-inch cable for the hitch and for connecting the feeder house chain to the pipe. This unit pulls easily behind an ATV.

–W.H., Nebraska

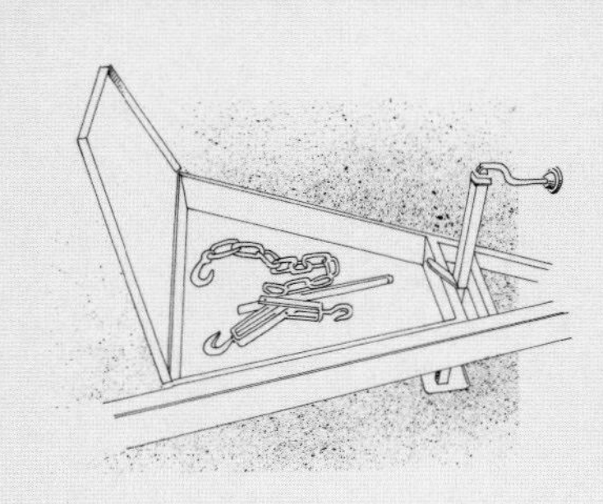

Trailer toolboxes

We made and mounted this toolbox to fit into the V-frame between the hitch and the bed of our 12,000-pound trailer, since the trailer didn't come with one. Now our log chains, binders, and pins are always right where we need them. It was inexpensive and it's out of the way.

–K.W., Illinois

Homemade mulcher

This mulcher for our seeding tractor is built of four three-bar, 6-foot mulchers. They are raised and lowered by cables when the three-point hitch is moved. The outside two mulchers fold up for transport. It helps to level the seedbed, and the wheat crop was our biggest in many years.

–W.R., Minnesota

Portable auger

We use this transport to move our auger. One person can clamp a trailer hitch to the end of the tube and tow the auger down the road. A chain keeps it from sliding along the frame in transit.

This way we don't have to have an auger and motor at each of our multiple bins in multiple locations.

–C.O., Minnesota

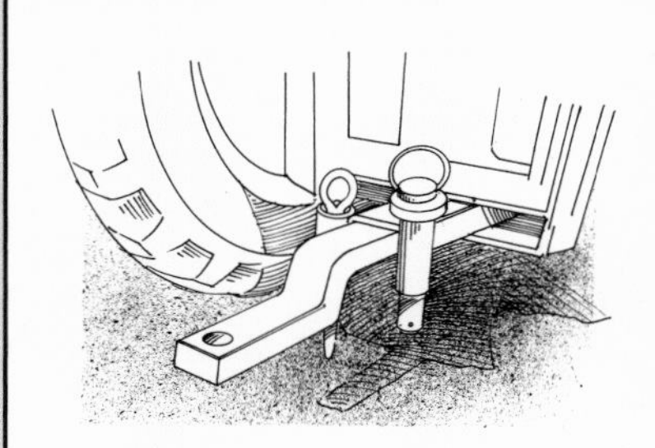

Larger storage holes

Since the storage holes in my tractor's drawbar are too small for the larger pins necessary for bigger machinery, I welded a 1¼x3-inch pipe to each end of a ¼x1-inch flat iron the width of my drawbar and then drilled a hole in it to bolt to the drawbar.

This accommodates the larger pins.

–D.V., Iowa

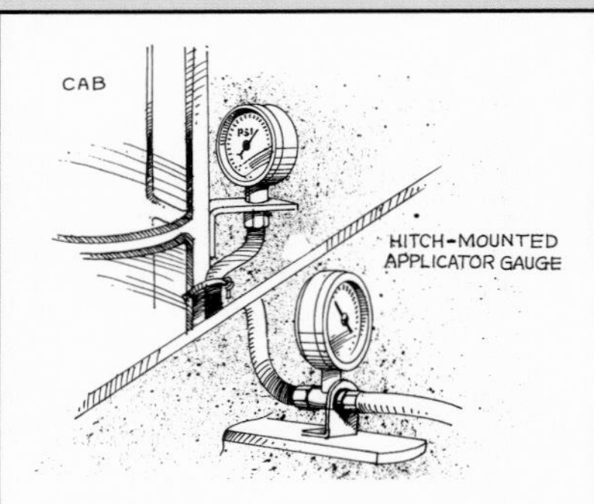

Anhydrous applicator add-on

Monitoring NH3 pressure is easier since I mounted another pressure gauge outside my front cab window. Using a ¼-inch pipe tee and nipples, I plumbed from the hitch-mounted applicator gauge and ran some NH3 hose around to the front of the cab. It's secured and fastened to a small bracket on the opposite side of where I get in and out of the cab.

–M.W., Kansas

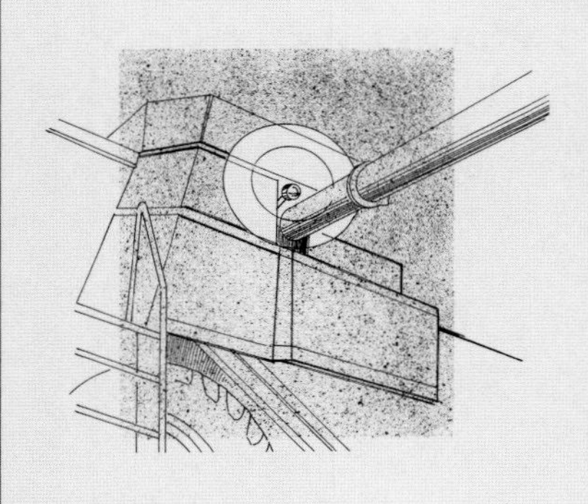

Unload auger indicator

The unload auger indicator light I installed under my hopper topper is wired to the electric unload switch. When unloading on-the-go, the auger cart driver knows instantly whether the auger is on or off without having to crane around to see. Also helps prevent auger or cart damage, particularly in rolling ground.

–K.H., Iowa

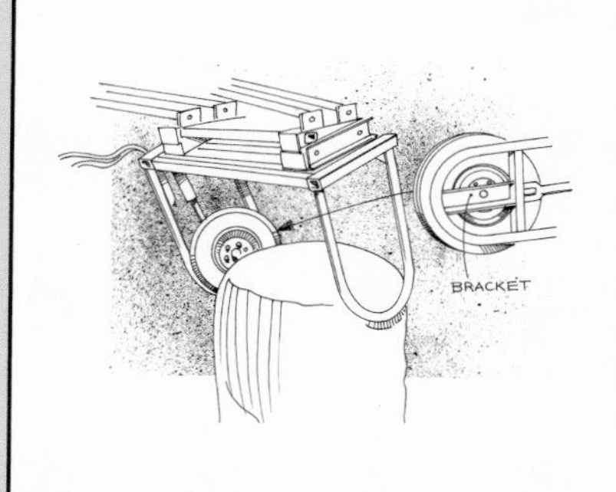

Bale grabber

To improve the holding power of the bale grabber mounted on our tractor's front end loader, I welded a bracket to the hydraulically moveable arm and bolted on a 14-inch tire and rim. The rubber sidewalls greatly reduce slippage on wet, snowy, or ice-covered plastic wraps.

–G.G., Massachusetts

Sprayer from a rotary hoe

I added on to the ends of the wings of a 31-foot flat-fold rotary hoe and made a 45-foot sprayer. The sprayer folds over the center section of the toolbar for narrow (16-foot-wide) transport. Castering wheels help with spray height control. The 3-foot ends break away if struck.

–K.P., Iowa

Stay out of the furrow

We take the sway blocks off the three-point hitch for our switch plow and pull out the pins to let the arm float up and down (about 3 inches). This lets it pull a lot straighter and makes it easier to stay close to the furrow without falling in.

–Posted by dutch

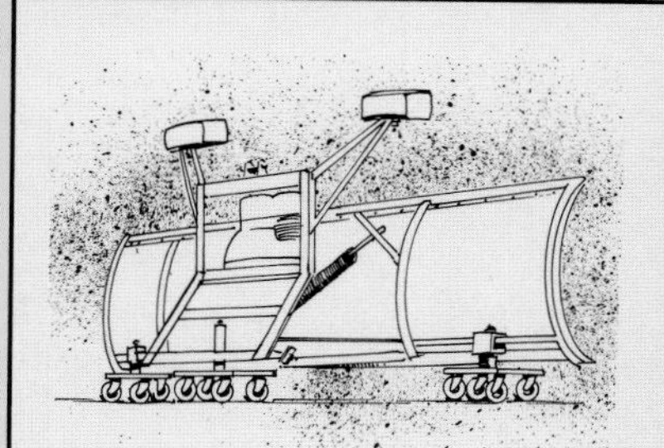

No more struggling with bulky snow blades

Mounting and dismounting a bulky snow blade is difficult at best. Being sure the blade is perfectly aligned when you get close is a problem. I mounted receivers on both the blade and undercarriage mounting brackets. I built a dolly with four dolly wheels and a spindle stud in the center.

Once in place and mounted, the blade can be raised and the dollys removed.

–R.C., Minnesota

Attention-getting skid steer horn

It's hard to hear, or be heard, in the cab of a skid steer. I mounted an old car horn and used a starter button for the switch. Now I can get a person's attention. The cows will move out of the way, too.

–D.M., Wisconsin

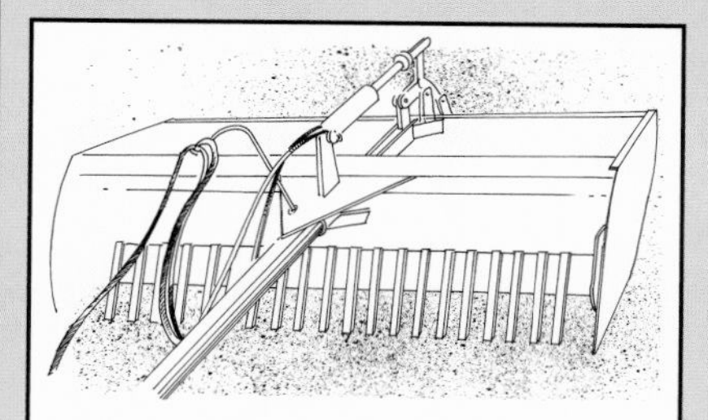

Homebuilt stone rake

I welded 1x1½-inch-thick bars spaced 4 inches on center to a length of 6-inch channel iron. I hooked the channel over the edge of the scraper blade and bolted through the holes holding the blade. Extending the teeth 9 inches below channel lets dirt filter through while holding stones in the scraper.

–L.K., Ohio

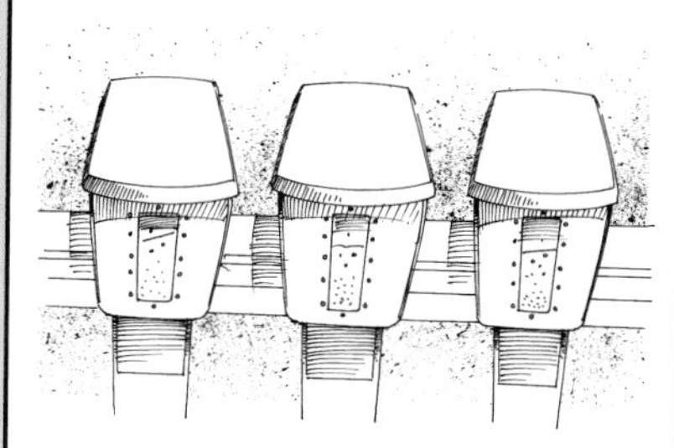

Planter box windows

It was difficult to see how much seed was left in my planter boxes without stopping the tractor. I cut 3-inch-wide, 12-inch-long pieces of clear, flat plastic and pop-riveted them to 1½-inch wide, 11-inch-long strips cut from each planter box. Now I can see, even with the lids on.

–K.Y, Missouri

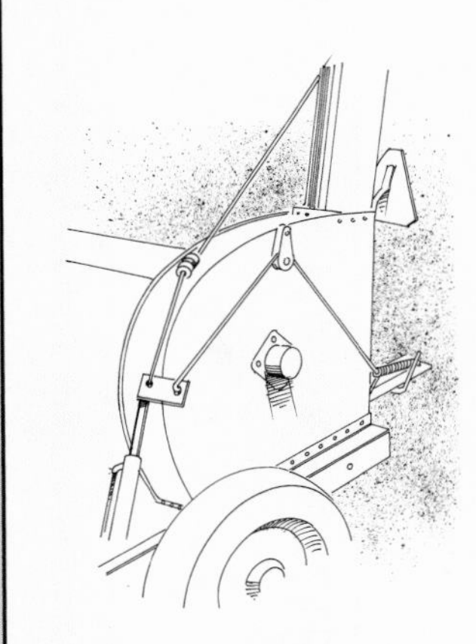

Unhooking forage wagons

What was a strenuous task is now as easy as flipping the toggle switch that operates the spout. I added a piece of 2-inch angle iron to the hydraulic cylinder that operates the spout. Then I attached a ¼-inch cable to the 2-inch angle iron. Contract the hydraulic cylinder and the cable pulls the quick-hitch latch, disconnecting the wagon. A spring in the cable system takes up tension if the cylinder is contracted too far. This takes about $5 in materials and a half hour to build.

–D.K., Illinois

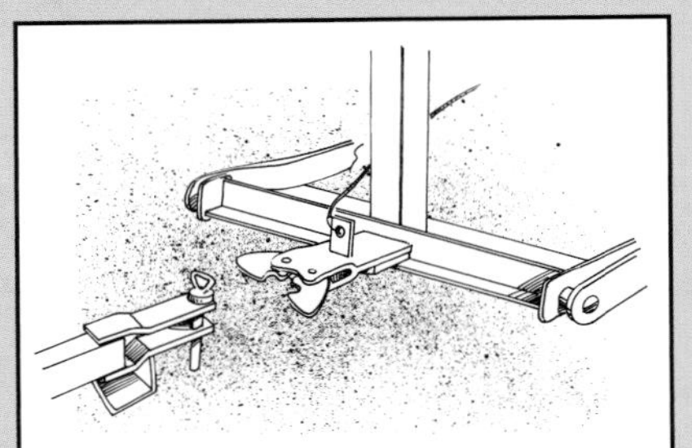

Faster hookups

This multipurpose three-point hitch is made from ¼-inch thick by 2x4-inch box iron. I welded ⅜-inch flat steel to the back side of the box iron with the hitch on the bottom.

When loading big bales, I bolt a quick hitch on so that I can load and move the wagons without having to get on and off the tractor. It also comes in handy for moving equipment.

–E.L., Ohio

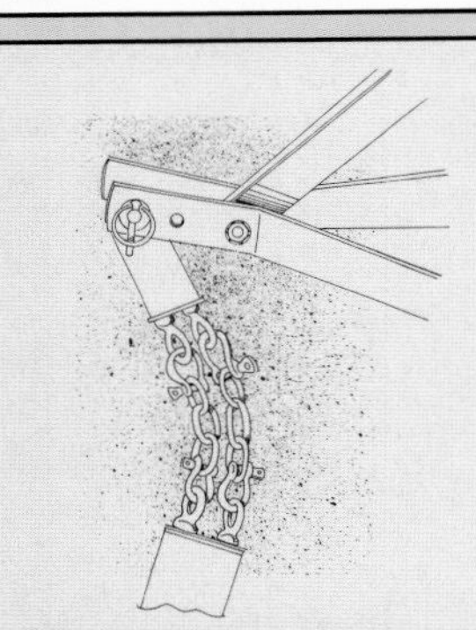

Modified bush cutter bracket

This device (two 1x3-inch scrap metal pieces, four ½-inch eyebolts, two pieces of chain, and four ⅜-inch shackles) replaces the top link of a three-point hitch to prevent the trailing wheel bracket on the back of our bush cutter from bending. It allows the cutter to rest on the lift arms.

–K.I., South Carolina

Fewer steps

We use this rack for unhooking a three-point hitch from the tractor and then for storing the hitch when it's not in use right on the rack. The wheels are important because they let us move the rack and the hitch out of our way.

–D.W., South Dakota

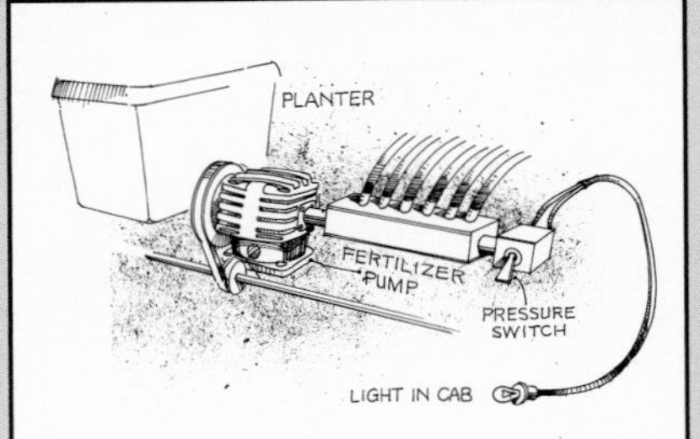

Pump pressure monitor

A small clearance light in my tractor cab is wired to an irrigation pressure switch on the output manifold of the fertilizer pump on my planter. When the pump is running, the light is out. If something goes wrong, the light comes on. Now I can tell if the pump is still running without looking back.

–R.R., Nebraska

Easy retrieval

I took a 6-inch piece of 1-inch PVC and glued an end cap to one end. The other end cap slides on and off. Then I mounted the holder to the trailer with mounting straps. Papers fit right inside.

–T.W., South Dakota

Deep snow marker

Last summer I pounded a fence post in front of the auger. Now when I'm pushing snow, I can tell how far the auger sticks out. This way I prevent collisions with the skid steer.

–M.S., Minnesota

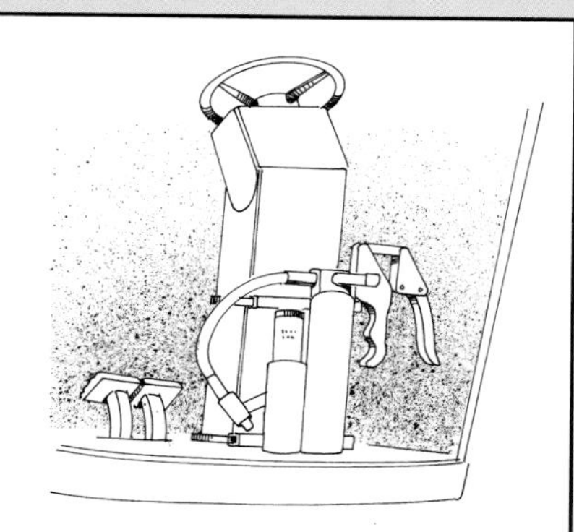

Grease gun holder

I built this holder for the grease gun and one extra tube with a tip holster on the side. It's strapped to the bottom, stationary, vertical portion of my combine's steering column with heavy-duty nylon straps. The gun stays clean and warm and out of the way.

–K.M., Illinois

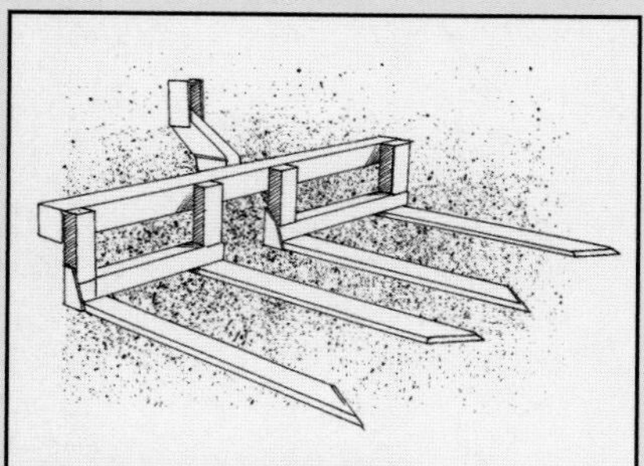

Double forklift

With a planter tool bar I got from another farmer and tubing I bought at a scrap yard, I made this implement. The forks are welded to the tool bar with adequate spacing for two rolls of hay to move right into storage. The top hitch is extended for adequate ground clearance on the rear. I shortened the tool bar and closed up the ends with flat metal.

–N.C., Mississippi

Versatile, flashing LED bicycle light

The handiest nighttime tool in my tractor toolbox is one of these lights. I can put one at the far end of the field to begin the field perfectly straight. They're also good for marking field drains or other obstacles so they don't sneak up, and they can also attach onto grain wagon frames.

–Posted by vanmaar

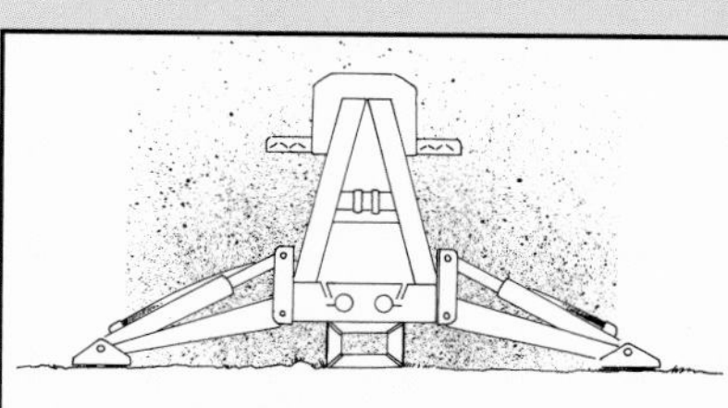

Backhoe support

My three-point backhoe sank down into the soil making it difficult to hook up again. I welded four pieces of 6-inch channel iron into a rectangular support about 12 inches wide, 8 inches high. Now the backhoe's lower center frame sits on the support, taking the weight off the stabilizer arms.

–J.K., Ohio

Feeding aid for corn header

I put ¾- to 1-inch lock washers on the outer edges of my auger. Wearing safety glasses, I used a hammer to drive the washers into place about 18 to 24 inches apart around the edges of the auger. They stay in place and help with feeding crop to the feeder house.

–Posted by kufarmerrancher

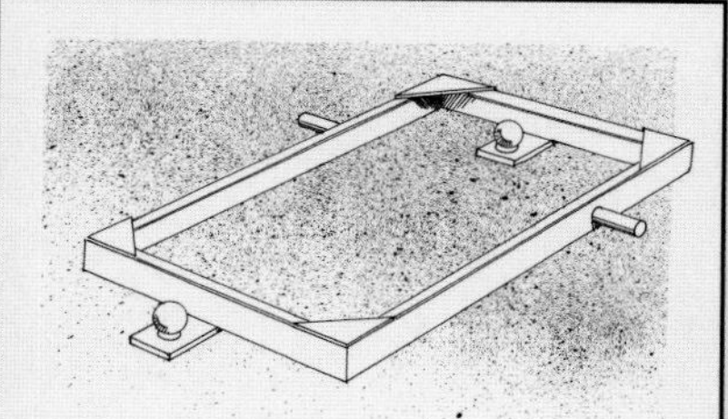

Homemade three-point adjustable hitch

I welded ¼-inch 2x4 metal tubing into a hitch. By mounting it vertically on my tractor, I use it to back a trailer or chute and still run the PTO shaft through it. We use receiver hitches. By pulling a rope we can hook or unhook without leaving the tractor. The quick hitch will lift high enough to use the tractor drawbar or low enough to set the quick hitch on the ground and unhook off the tractor without tipping over.

–D.Z., South Dakota

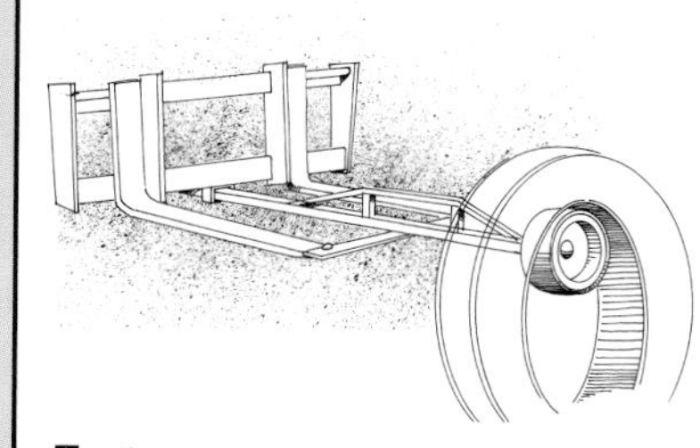

Extra arms

We built this convenient but inexpensive boom for mounting and removing axle-mounted tractor dual wheels. The trussed bar with a cultivator wheel and spindle makes mounting it onto the loader forks a quick job. The dual can then be rotated easily when it's time to install bolts.

–B.S., Iowa

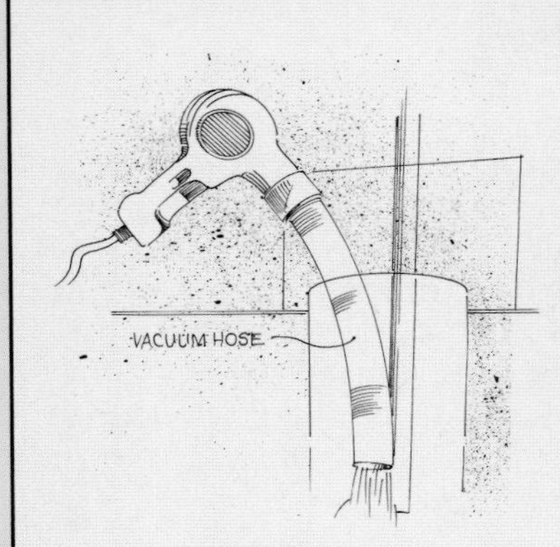

Handy thawing device

When we have a water pipe that freezes down in a tile, we often take a shop vac hose or similar tubing and feed it down the tile line hooking the other end to a blow drier to get the heat to the bottom of the line. It gives you more flexibility of getting the heat where it belongs. This is much safer and does much less damage than using a torch.

–W.A., Iowa

Guard against breakage

To prevent the stems from breaking off tubeless tires, weld 1-inch pieces of 2-inch round tubing to the rim around the stem. It makes a simple, inexpensive guard against flying chunks of dirt.

–J.W., South Dakota

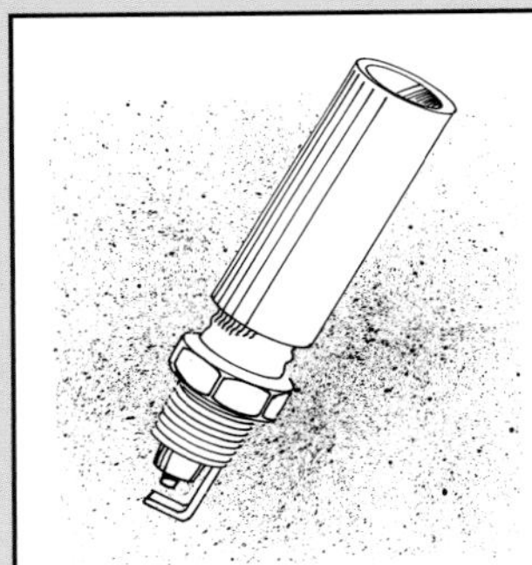

For hard-to-get-to plugs

There are times when it's hard to hand start a spark plug, especially where the space is limited. I use a 5-inch-long piece of air hose with an inner diameter of ½ inch. The spark plug fits perfectly. It's neck slides into the hose. Try it – it works great.

–J.T., Michigan

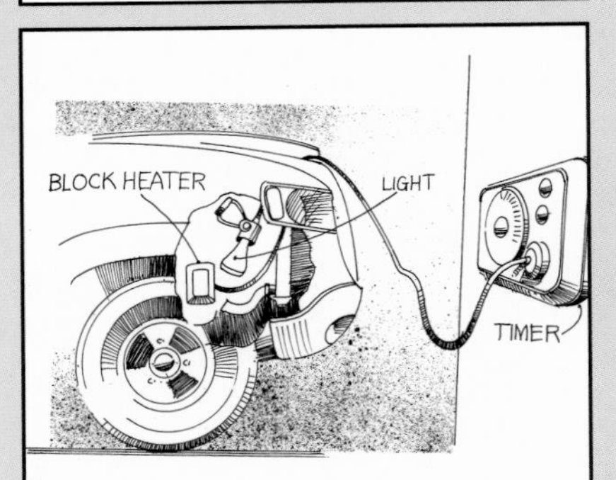

Cold-weather energy saver for vehicles

We have used a timer on our engine heaters to save energy costs yet still give our vehicle engines ample time to heat up.

We added one feature to give us confidence at a glance that the timer has kicked in and that everything is working. We wired a small electric light in parallel with the circuit to the engine heater. When we see the light is on, we know the timer has kicked in.

–B.B., North Dakota

A tip to avoid rusty bolts

Bolts around exhaust manifolds or superchargers get hot and tend to rust. The same is true for battery cable ends or grain drill seed adjustments. By replacing these with stainless steel hardware and/or wing nuts that will not corrode, I don't have to have a wrench to open them.

–J.K., Ohio

Shorter belt-replacement time

Put new replacement belts on, over, and past the existing ones, and cable-tie them to the guard or combine panel. That way, when the existing one breaks, you can cut the cable tie, slide the new belt on, and keep on harvesting with only minimum downtime.

–Posted by Brad

Unstick stubborn axle hubs

We had trouble removing stuck hubs for years. Finally we found a heavy pipe that would slip over the axle, then cut a part of it off so it will only hit the wedge. Then we welded a heavy square shaft for something to hit. A couple good strong hits will jar it loose.

–Posted by ralph

Clearing a clogged pipe

To clear a clogged piece of hose or pipe, I attach a plastic bag to one end of the pipe and tie it securely with a piece of twine. This creates a positive vacuum.

On the opposite end, I connect a shop vac. With the bag in place, I have cleared sections of pipe up to 1,000 feet in length with no problem. Both ends must be securely sealed for this to work effectively.

–T.W., South Dakota

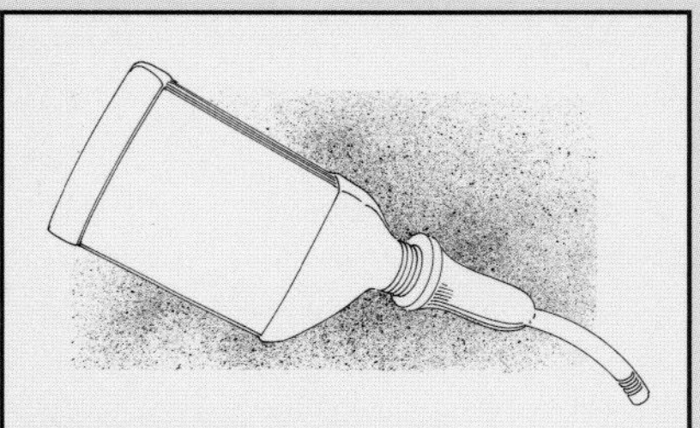

No-drip oil can

I found that a used milker inflation tube will fit snugly over a 1-quart plastic oil can and makes a pour spout for hard-to-reach transmissions on combines.

The milker insertions are flexible, can be bent and inserted in narrow areas, and fit into side filler holes that are so hard to get to.

–T.B., Illinois

Portable 100-psi air tank

My broken air compressor is back in use. I put a ½-inch tee with a pressure gauge on the top. On the side outlet of the tee, I installed a ½-inch ball valve and an air hose nipple that plugs into my shop air hose connector. An 8-foot air hose goes to the tank outlet with a tire chuck on the end.

–R.C., Michigan

Keep tires in shape by adding strength

We've had a hard time keeping rims round and the tires aired up on skid loaders. To correct the problem, we rolled ½x1-inch bar stock then welded it to the outer rim bead. This saves a lot of downtime.

–J.K., New York

Quick, safe answer to cold problem

Caught with frozen water lines, I grabbed a 2-quart pump-type sprayer and filled it with hot water. With the nozzle turned to the stream setting, I had a safe, effective, and quick solution to my problem.

–R.K., Wisconsin

Simple solution

If you are ever faced with locking two nuts together that must stay fixed, we solved this problem by placing a lock washer between the two nuts. This technique has never failed me, nor have I ever had a bolt come loose.

–S.A, Nebraska

Removing worn bushings

To dislodge worn bushings, center a steel ball with locking pliers over the center of the bearing. The ball will automatically center itself in the bearing. You can beat on the ball as hard as you want until it falls out. If the ball is the same size as the housing, it will push it on out without scoring or damaging the race in the process.

–J.B., Illinois

Hands-free painting

If you're using an aluminum ladder to paint a building, you may already have a paint can holder right in your kitchen. Tired of holding the heavy can with one hand, I got a long-handled 8-inch pot, stuck it in the end of the rung on the ladder, and put the paint can in the pot. Of course, the handle fits anywhere up or down the ladder and on either side.

–D.R., Iowa

Handy nail and screw holder

I took the insides out of an old wristwatch and welded a flat cow magnet to it. Now I carry nails and screws on this wristband with a magnet instead of in my mouth.

–B.Q., Wisconsin

Custom-made oil seal installation tool

To fit a 3¼-inch oil seal, I welded together a ¾-inch nut, a ¾x5-inch machine bolt with the head cut off, and several very large flat washers, smoothing the bottom washer with my lathe. Works nice.

–J.K., Ohio

Easier cleanup for riding mower deck

While I had the deck off of my riding mower, I sandblasted and painted the topside. Then I applied two coats of fiberglass resin to the underside. It's easier to wash now, and the coating will last longer than paint.

– Posted by sathomsen1

Foolproof oil change

When you drain the crankcase on any tractor, truck, or car, put a wrench on the seat. Leave it there until the new oil filter and oil are in. This simple reminder might save an engine overhaul.

–C.P., Illinois

Multipurpose mirror

We duct-taped a motorcycle mirror to a ½-inchx4-foot fiberglass fence post to look for wasp nests under farm equipment. We also use it to look inside machinery during repairs.

–R.H., Washington

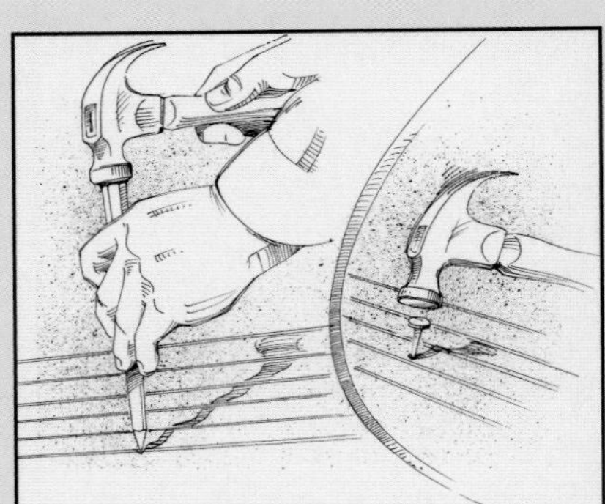

Galvanized metal projects

I sharpened this ¼x12-inch nail punch to a 20°-angle point. I use it for starting roofing nails when applying galvanized metal to buildings and roofs. It saves me time, not to mention painful injuries to thumbnails and knuckles.

Nail holes can be enlarged as needed.

–B.S., Missouri

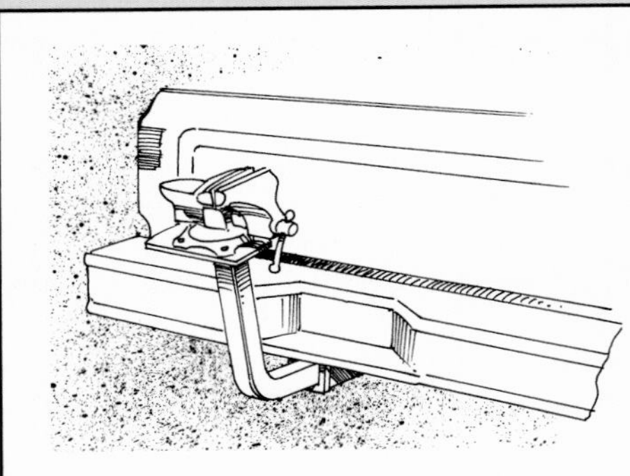

Portable vise holder

I welded a bracket with a pad for bolting a vise to the square tubing that fits into the receiver hitch on the rear of my pickup. The mounting arm can be whatever length I decide on so the height will be comfortable.

It works very well when I'm in the field for sharpening my chain saw or for working on other machinery or equipment.

–T.H., Oklahoma

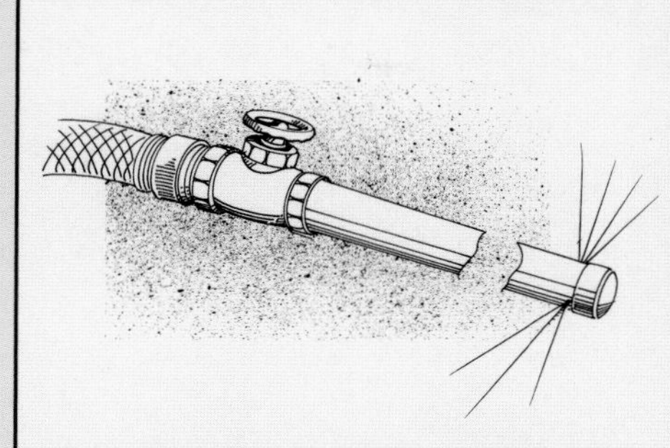

Radiator washing wand

I made this wand from a 5-foot length of ½-inch rigid copper tubing with a sweating cap on one end and a valve and female garden hose adapter on the other. There is an ⅛-inch hole drilled ½ inch from the capped end. The valve controls the water and makes a handle, too. It took about 30 minutes to make and cost $13.45.

–J.P., Kansas

Winterizing disks

I cut a 30-gallon tank in half and mounted a wheel kit to the base so it can roll under my disk. Filled ¾ full with waste oil, I slide it under a gang, then lower and spin the gang until all area is covered with a thin layer of oil. Two people can do a complete disk in less than 10 minutes.

–K.C., Illinois

Back in action

When the stock handle on our top link was no longer able to adjust the work angle of a very heavy implement, I had a machine shop bore out a large nut to the outside diameter of the turnbuckle.

Now I just keep the correct size wrench in the toolbox.

–S.J., Tennessee

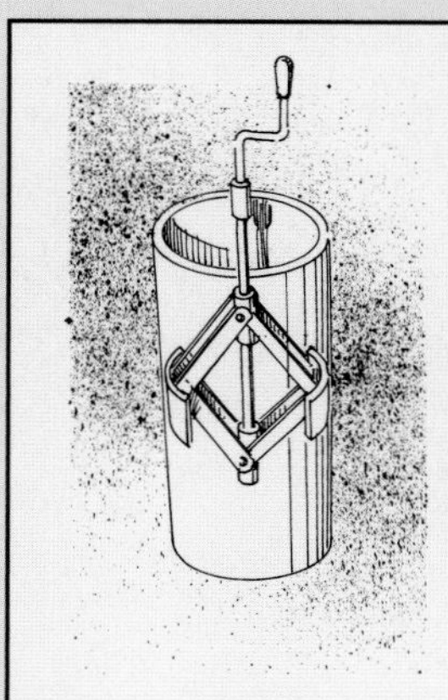

Scissors jack makes an excellent repair tool

I found that a scissors jack from a junk car makes an excellent tool to remove or repair dents in damaged auger tubes.

I have adapted a couple of jacks for various-diameter tubes. A radial section of heavy tube welded to both sides of the jack acts to push out dents. A shaft or tube welded to a screw helps to navigate the jack inside the tube as well as to tighten the screw.

–R.D., Michigan

What's the best way to sharpen a chain saw?

We have a brand-name sharpener, but all you need is any high-speed rotary tool that takes a ⅛" shaft. Then just get the right size stone. I threw out the guides that came with ours and do it freehand. Works great!

– Posted by crossman

Replacement beater substitute

Our manure spreader needed a new beater. We used a 20-inch auger with ¼x1½x4-inch pieces of metal welded 6 inches apart on the flighting. It works well and was cheaper than brand-new.

–V. S., Pennsylvania

There's more than one way

It's tricky to replace hard-to-reach screws; they fall right out of the socket. Try cementing the screw head to the socket with a small dab of five-minute epoxy glue. Then break the socket loose after tightening the screw, when the epoxy dries and becomes brittle.

–J.K., Ohio

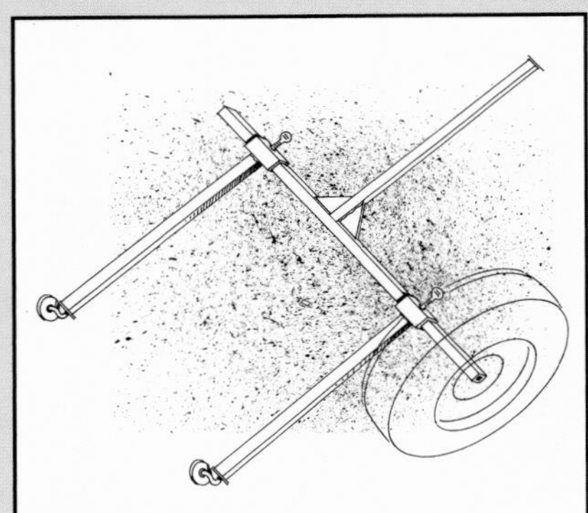

Move big tires more easily

I move my truck and front-wheel-drive tractor tires around on this device. It is constructed of 2-inch square tubing with small pieces of 2³⁄₁₆-inch tubing that support the front arms. The arms also adjust to different tire sizes by sliding over the 2-inch tubing.

When I use the third extension as a lever, the tire and rim lift up right to the height of the hub.

–K.U, Texas

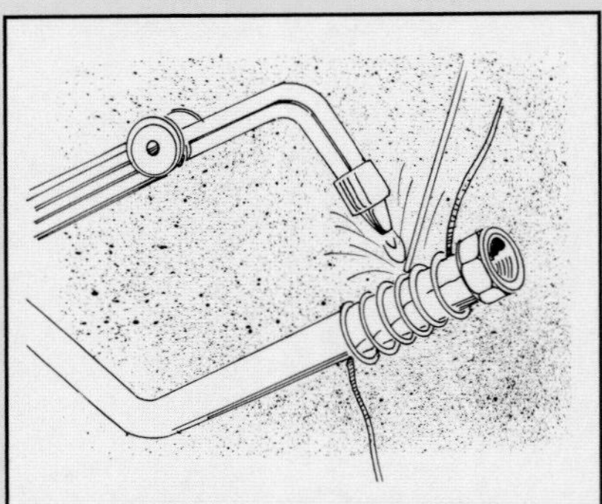

Easy repair for steel hydraulic lines

When repairing split-steel hydraulic lines I first wrap a coil of electric fence wiring around the line. I leave a ¹⁄₁₆- to ⅛-inch space between the wrapped coils. The wire will give the repair added strength as well as help hold a thicker layer of brazing rod.

This technique has always given us a permanent repair even under pressure.

–K.M., Indiana

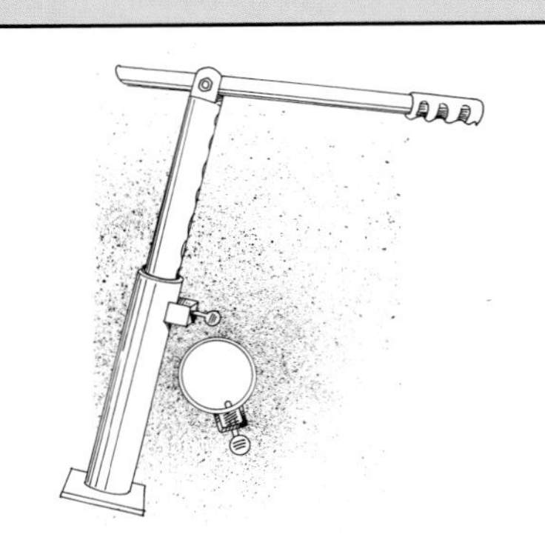

Check bearing stability

I made an adjustable prying tool to check my bearings or pulley shafts on my combine before harvest season to determine if any were worn or needed to be replaced.

By adjusting the height of the tool, I can pry upward on the outside of the shaft to check for even very slight wear or movement within the bearing. It doesn't guarantee a trouble-free harvest, but it certainly helps start out with no problems.

–J.K., Ohio

Easier jack retrieval

To avoid crawling under my machinery, I lengthen the handle on my hydraulic jack by 12 to 18 inches with round stock or pipe. A 6- to 8-inch T-handle at the end made from the pipe has curved ends.

–K.U., Texas

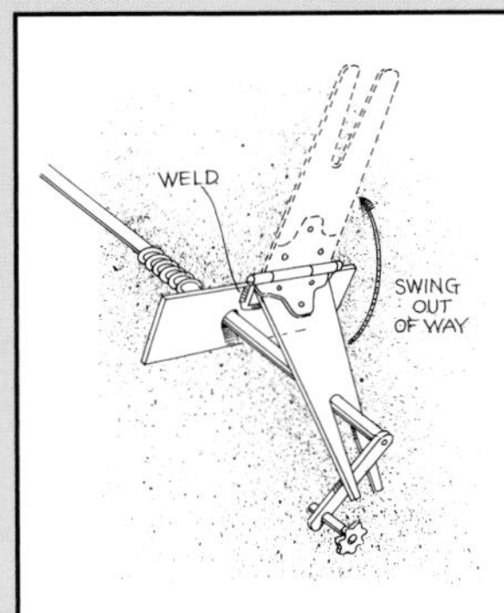

Fast lock for hay rake

The adjusting arm on my side delivery rake kept shaking and vibrating until it unscrewed itself. Then the rake teeth would drag the ground and wear out. So I made a locking adjusting arm with a door hinge welded to the rake at one end and a two-prong fork on the other. The fork is long enough to lock the arm at its elbow.

–H.P., Georgia

Reincarnated bicycle spokes

I needed something to clean out orifices for a carb rebuild, but wire wasn't rigid enough. Then I spied a discarded bike wheel. Cutting near the axle hub and slid out one of the spokes, leaving the nipple on for a handle. Spokes are bendable, disposable, and can be cut to any length.

–Posted by lloyd

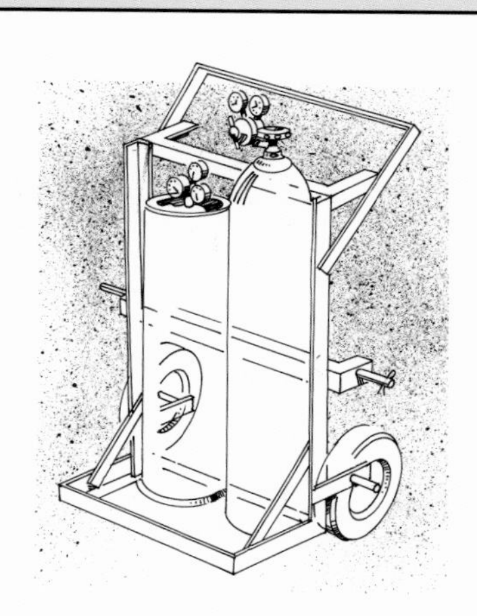

A three-point hitch for acetylene cart

We add a three-point hitch to our oxygen acetylene welder torch cart for both safety and convenience of transport. The handle and toolbox on top allows for tool storage. A top link was added just below the toolbox. A second bracket just above the wheels has two pickup pins for the tractor three point. An additional bar was added at the bottom to keep tanks stable and from sliding sideways during transport from farm to farm. We use a 30-inch cart for more stability.

–B.S., Missouri

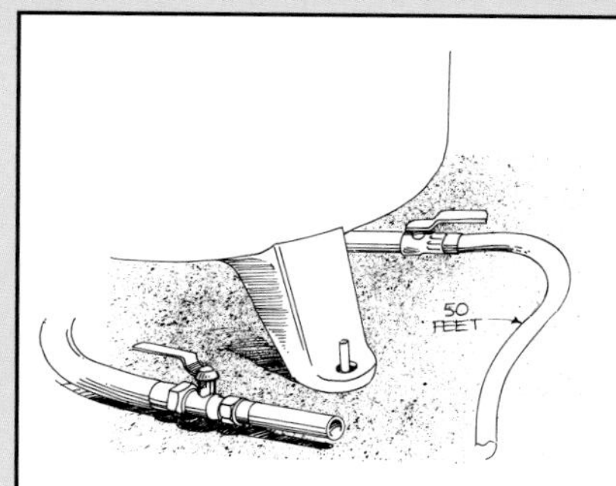

One machine, more uses

We added two ½-inch gate valves to our 60-gallon compressor, one with 50 feet of air hose, and one with 6 inches of ¼-inch air hose. After draining the water that collects in the tank, we use the hose to clean machinery with the big blast of air. Very helpful for mounting stubborn tires, too.

–G.S., Nebraska

Abrasive cloth roll holder

There are four rolls of 1-inch-wide cloth-backed abrasive mounted on a ⅝-inch-diameter wooden dowel in my shop. The cloth is very handy for cleaning rust off machinery or for polishing parts.

–J.K., Ohio

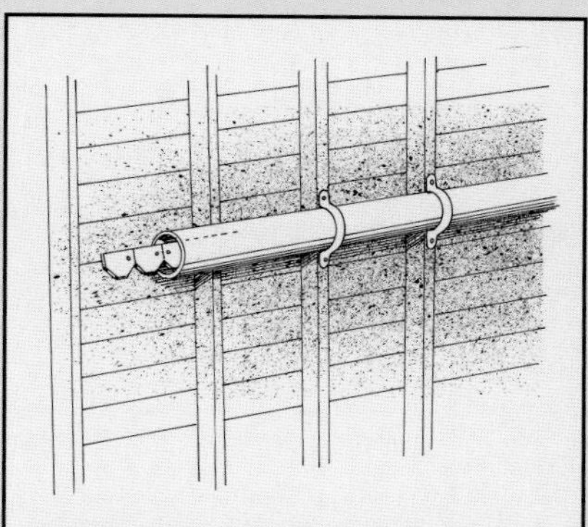

Safe storage on the machine shed wall

To store our extra 18-foot-long combine knives, we mounted a piece of 3-inch PVC pipe to hold them. It is attached to the inside wall of our machinery shed near the big door. We simply slide the knives in for safe storage when they're not in use. This system works for any length knives just by making the pipe longer or shorter.

–L.K., Ohio

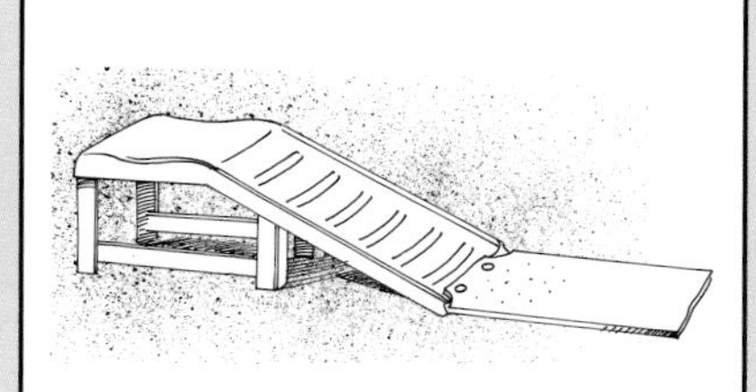

No more creeping ramps

A mud flap or piece of carpeting bolted to the front edge of the car ramp keeps the ramp from sliding forward on concrete garage floors. When you pull the vehicle forward onto the flap, whether it be two- or four-wheel drive, the weight of it keeps the ramp in place.

–J.N., Missouri

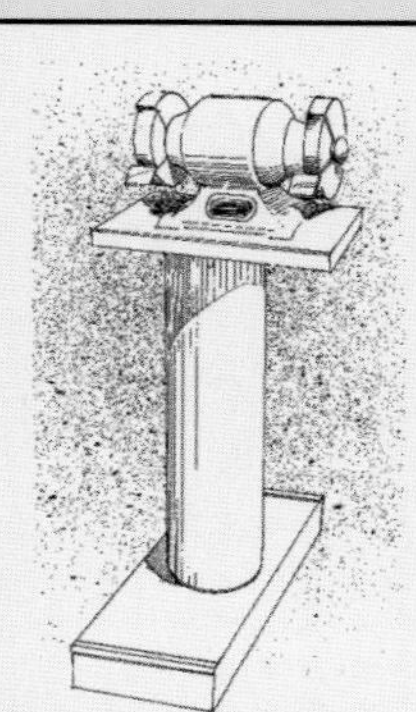

Moveable grinder stand

This stand is made of aluminum-wrapped 2x12s, 8-inch PVC well casing, and plastic laminate countertop. It's light enough to move around and stout enough to resist pressure applied during grinding. I got the stability I needed by orienting the base 90° opposed to the grinder mounting surface.

–T.F., Iowa

What tool is suggested for cutting 10-gauge steel?

For light to moderately heavy material, I use an electric nibbler. They run on 110 volts and have a steel bit that moves up and down and spits out bits ¼"×⅛". They usually leave a smooth finish that may be a little sharp, but won't cut you. They're a lot like a jigsaw to use.

– Posted by farmboy4

What's the best way to make T-joint cuts for welding gates?

I bought a quick and neat tubing notcher that makes perfect cuts every time. I can make 90s, 60s, 30s – almost any angle with the pull of a handle and faster than a chop saw. There are different sizes that cost around $250. I've had one for 10 years and made hundreds of gates.

– Posted by mayday

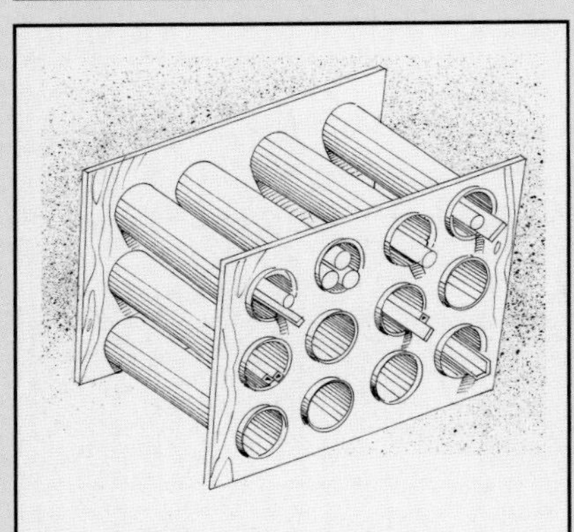

Steel storage rack

We took a 4x8 sheet of ¾-inch plywood and cut 6-inch holes in it. The holes need to be just large enough to accept 6-inch drain pipe, which we cut 10 feet in length (or whatever length you desire). We put the plywood on both ends to form a freestanding rack.

We store different dimensions of pipe or rod in each section, making it easy to sort out what we want for a particular job.

–C.M., Missouri

New battery charger cart

I stretched the seat belt of a used, $2 baby stroller and made it into a battery charger cart. I took the wheels off the charger and pulled the axle out. Bottle and diaper pockets hold terminal cleaners and post wrenches. Clamp the cables to the visor to prevent a short.

The wheel locks are useful in my hilly yard, which is easier to cross now with the big wheels.

–E.H., Wisconsin

Sandblast cabinet

I made this cabinet from a used 250-gallon fuel tank and some scrap steel that was lying around my farm. One switch turns on two floodlights and an exhaust fan. It rolls outside for blasting big items. I also use it as a paint booth – wet items dry on hooks attached to a bar running lengthwise near the top.

–R.D., Illinois

Free-swinging shop lights

We had some problems with shadows when working on motors in our shop. So we doubled our lights by adding some long tube lights hung from chains. And if we run into them, they swing.

–Posted by ph4

Connector caulking cap

When I have a partial tube of caulking left over in the caulking gun, I screw on a solderless electric connector over the tip. It does a good job of sealing the tube.

–W.G., Minnesota

Catch filings and oil

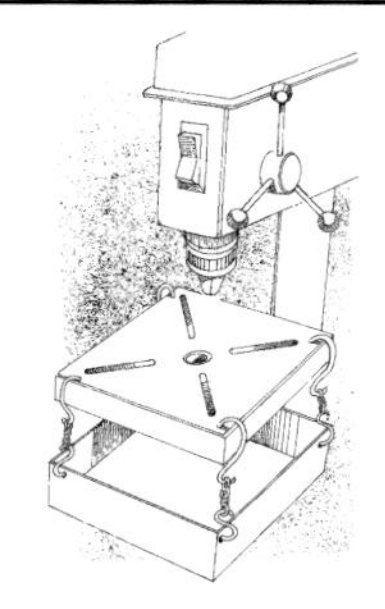

This steel box pan is suspended by S-hooks and small chain from each corner of my drill press table. I made the pan the same size as the table. It catches an amazing amount of metal filings and keeps the cutting oil from getting on the floor through the center hole.

–B.A., Montana

No more tangled straps

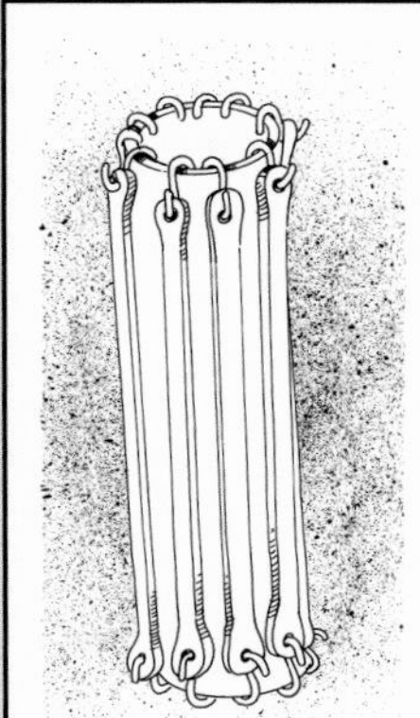

We use stiff cardboard tubing or sometimes if we are storing them outside we use PVC pipe to make a convenient, lightweight holder to keep tarp straps in order.

We cut the tubing approximately 4 inches longer than the extended tarp straps. Hook the straps to the tubing as shown. You can add notches in the tubing. With the straps on the outside of the tubing, this puts the open hooks safely inside where you can't get snagged on them.

–S.S., Ohio

Setting saver

When I use a locking C-clamp for repeated clamping, it sometimes tends to lose its setting. So to eliminate that problem, I thread an SAE 7⁄16-inch wing nut (or just a regular nut) on to the pliers' adjustment screw and then jam-lock it by countertorquing.

– Posted by jhofer

Extra storage space

Most tool cabinet manufacturers set the top cabinet directly on top of the bottom unit. I think this is a waste of shelf space. I built a spacer out of 6- to 8-inch channel iron to put on top of the bottom unit and set the top one on it. On the outside of the spacer, I bolt wide, lightweight angle iron onto it and paint the chests white to make the shop seem brighter.

–C.K., Iowa

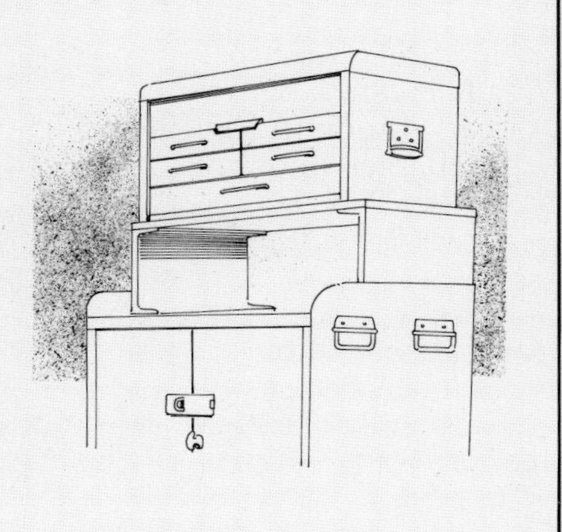

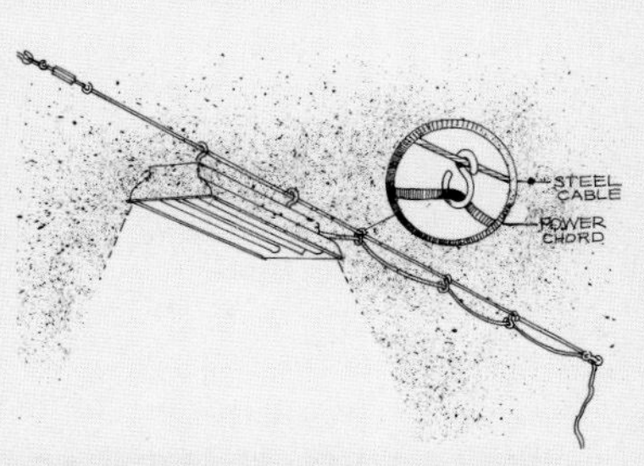

Portable shop light

This 4-foot box fluorescent light has two eyebolts on top and hangs on steel cable strung across my shop. The eyebolts slide on the wire, which has a turnbuckle on one end to keep it tight.

I looped the electric cord from one side of the shop to the other with S-hooks every 2 feet. The cord is taped to the S-hooks. Now I have light where I need it.

–F.S., Missouri

Traveling trouble light

We get more use out of our reel trouble light now that we mounted it to the magnet from a small grinder mixer. Since our shop walls are steel and tin, we can move it wherever we need.

–M.W., South Dakota

Snap hooks for holding wrenches

When I have to carry a lot of wrenches in an implement toolbox, I can find the one I need quickly by hooking them all on a couple of snap hooks. Sometimes I can use a wrench right on the hook. Otherwise, it's easy to slip it off.

–L.P., Minnesota

Tools stay in place

I cut pieces of tempered masonite to protect my toolbox drawers, but the tools slid around. So I coated the bottoms of the drawers with spray-on bed liner. Tools don't slide, and the bed liner preserves the masonite.

–M.E., Illinois

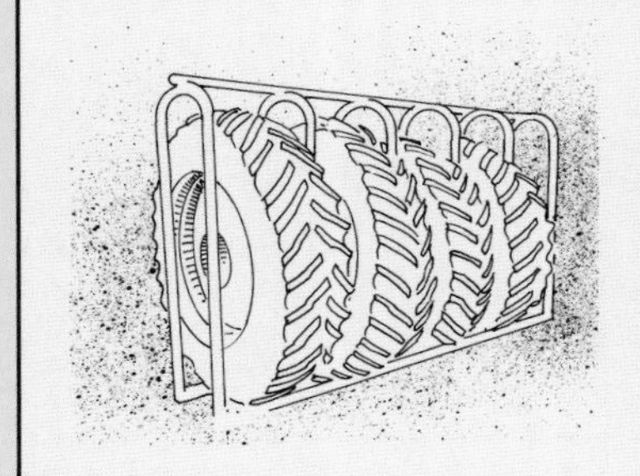

Dual tractor wheel stand

This stand is made of 2-inch pipe, and the dividers are spaced 2 feet apart. I built it primarily as a preventive safety measure. Since youngsters aren't able to roll the tires out of the stand, we can eliminate the risk of getting caught under a tire should it accidentally fall.

–L.T., Minnesota

How do I best install in-floor radiant heat?

In 19 years I haven't spent one dime on maintenance of pipes or pumps. We used rolled black plastic with a clear core and put the pipes 24 inches apart.

The 1/25-hp., 110-volt motor runs 24 hours a day and puts 2 pounds of pressure on the system.

–Posted by bigM

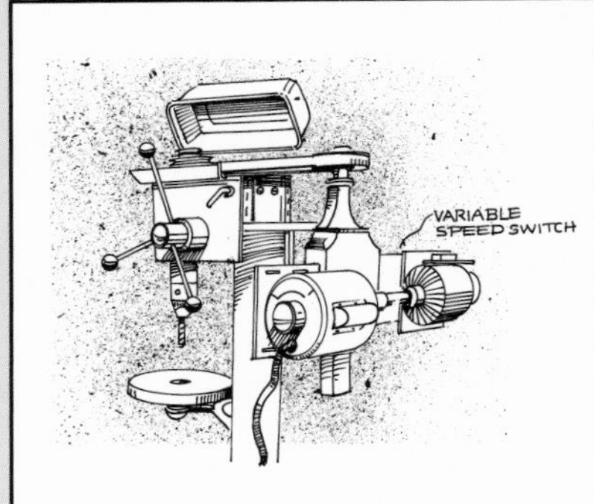

Saves time and drill bits

My drill press ran too fast for large bits and too slow for small ones. I didn't always take the time to change belts and, as a result, would ruin bits. So I used the hydrostat from a lawn tractor to power my drill. Now I can go from 0 to 450 rpm by simply moving a lever.

–R.G., Kansas

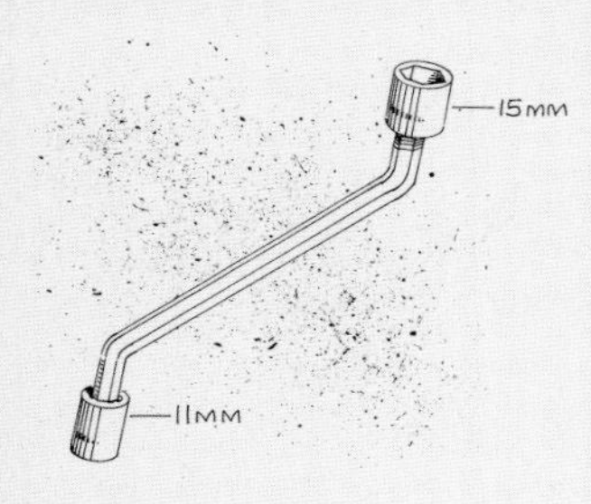

Homemade wrench

This wrench stays on my combine. I use the 11-millimeter socket for section bolts and the 15-millimeter socket works on guard bolts. The handle is made from a 12-inch piece of 3/8-inch keystock. The 3/8-inch drives are welded on. It comes in handy for making sickle repairs.

–R.R., Missouri

Fuel tank nozzle caps

Keep water, dirt, and spider webs out of fuel tank nozzles and gas can spouts by using cane tips or rubber tips from chair legs. They can be purchased in various sizes to get the right fit.

–J.H., Missouri

Lightweight dolly

I reinforced the bottom of an old power push mower handle and bolted the wheels to the handle's mounting holes. A tray welded between the handle and a brace is nice. I use it for moving around my battery charger or a small welder. I recycled the mower's deck as scrap iron and gave the old engine to a grateful high school shop class!

–J.J., Minnesota

Under-the-hood illumination

Even in a well-lit shop, engine work can be dark. A 4-foot fluorescent light fixture screwed to a 2x4 and mounted with strap iron and a pin to our engine hoist makes an easily removed light that hangs on the wall when not in use and adjusts to any height.

–J.C., Iowa

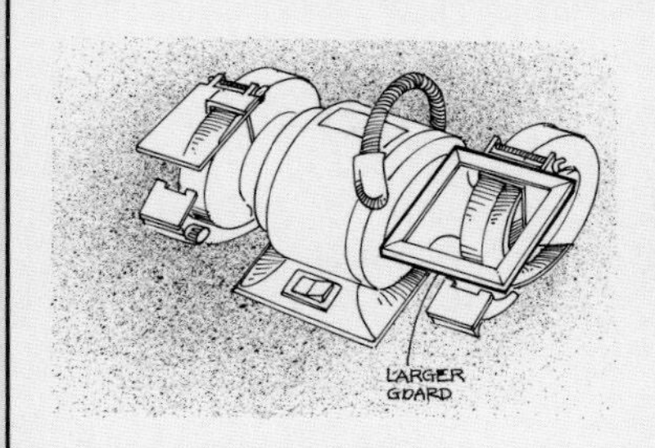

Larger shields protect eyes

The 2x4-inch safety shields on our new bench grinder were too small, so we put a 5x5-inch piece of shatter-proof safety glass in a 5¼x5¼-inch frame.

A U-shape bracket made of ½x⅛-inch flat iron allows frontal adjustment and attaches to the grinder wheel guard bracket.

–B.S., Missouri

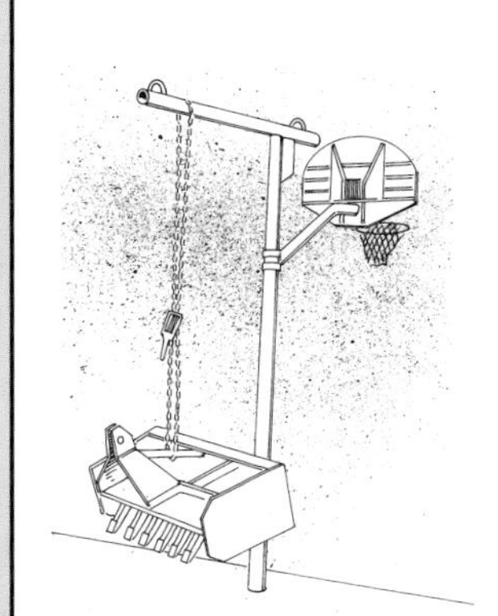

Basketball post hoist

I used 5½-inch od (outside diameter) well casing for our basketball goalpost because I had some handy. Taking advantage then of its strength and rigidity, I welded a 5-foot-long arm of 2⅞-inch od drilling pipe to the top of the casing post as a cantilever arm for hoisting; 1 foot of the arm extends from the post to the rear to provide a substantial gusset. It will hold a 1,000-pound load. The 5½-inch casing post is cemented 6 feet into the ground and reaches 13 feet high.

–D.C., Texas

Helpmate handle for air compressors

Lifting my 5.5-horsepower, 30-gallon air compressor over the ledge at our shop or onto a truck was a job. There was nowhere to grab it to lift it.

This prompted me to make a simple tool to slip around the tank to lift it without trouble. I used ⅜-inch rod, and I used ¾-inch pipe for the handle.

–M.E., Illinois

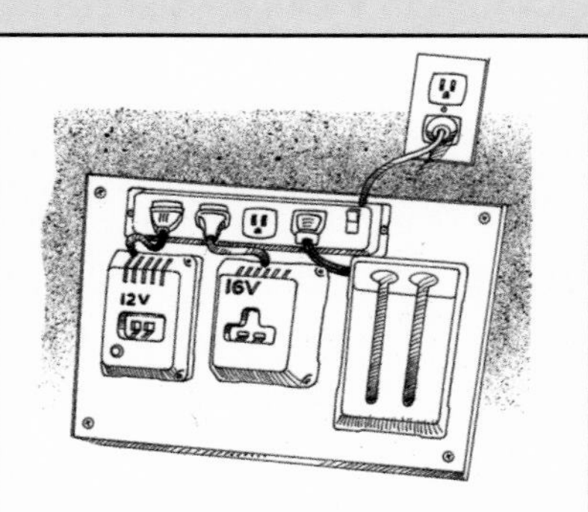

Consolidating chargers

Tired of having the chargers for our electrical tools all in different places, we mounted them on a ½-inch piece of plywood. We also mounted a power strip on the plywood to plug the tools into. The plywood hangs in a handy location in our shop so we have the batteries when we need them.

–B.M., Illinois

How do I weld large pieces of metal without a bow?

Tack-weld it all securely in position. Don't just start at one end and weld a continuous bead – break it up into sections and move around while welding. Don't rush the job.

If you get the whole thing hot enough it will warp no matter how much jumping around you've done.

–Posted by haywiremechanic

Work stations on wheels

I have more of these than I can count. The biggest is 6x8 feet, on a pair of 14-inch tires, with jacklegs on all corners to level it. Building or repairing machines on a cart sure beats working on your knees. Being able to move a project out the door while you're waiting for parts is good, too.

–Posted by Franz

Safety first fieldwork

We need to be conscious of taking a break from long, hard hours in the field, so I made a large white flag with the letters "Break Time" in red. I attach it to a flag pole that can be stuck into the pickup box stake holders. Every mid-morning and afternoon I place the truck at the end of the field so everyone will know it is time for a snack and rest.

–V.T., Iowa

A view of the truck bed

I mounted a convex mirror on a length of ½-inch steel conduit. It goes through a hole I drilled in the top of my truck bumper. A bolt through the bottom of the bumper holds it in place during loading. I can see the trailer in the mirror from my cab and know when to move the truck.

– Posted by Jeff

Two-in-one hitch

I put both a ball and a pin hitch on my new pickup in the same unit, then I can use either one that is appropriate. I welded a receiver crosswise below the regular receiver so as to store it out of the way when not in use.

–B.K.,Oklahoma

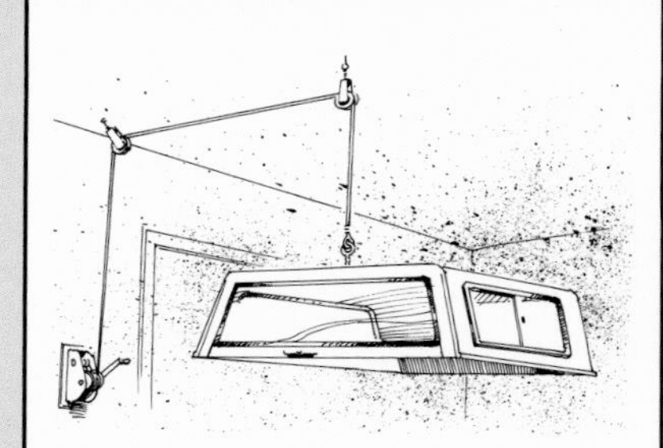

One-person topper hanger

It used to take four people to put a pickup topper in place or take it off. That many people were not always available. We took a ⅜-inch I-bolt and placed it perfectly in the center in the top of the topper. Then we placed a hand winch on the wall and connected the cable through two pulleys, one directly over the topper.

Now one person can do this alone. It's especially important if you have a gooseneck hitch in the bed of the truck that you need to get to and likewise sometimes need the topper cover.

–R.H., South Dakota

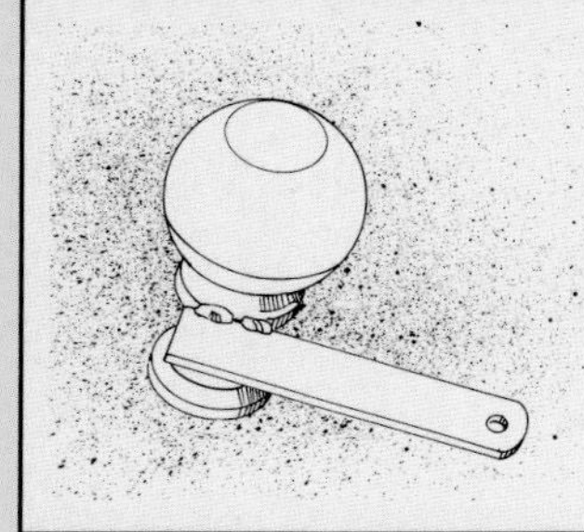

Quick-change ball hitches

Many times it takes two wrenches to change balls on a bumper hitch. To keep the ball from turning while I am removing the nut at the bottom, I welded a short piece of flat iron that acts as a handle at the base of the ball.

–R.K., Kansas

Better lighting for truck parking

It can be tricky backing a truck into a dark building. So I put two strips of reflective tape (the kind used on sides of trucks and trailers) on the back wall. The taillights put out plenty of light to see by.

–E.M., Illinois

No more frosty windshields

Suspend a rope or a 2x4 between two posts, far enough apart to drive between them and up about 6 inches higher than the top of the car. Attach a tarp to the rope or 2x4. Drive into the tarp, and it will flow over the car.

–L.G., Montana

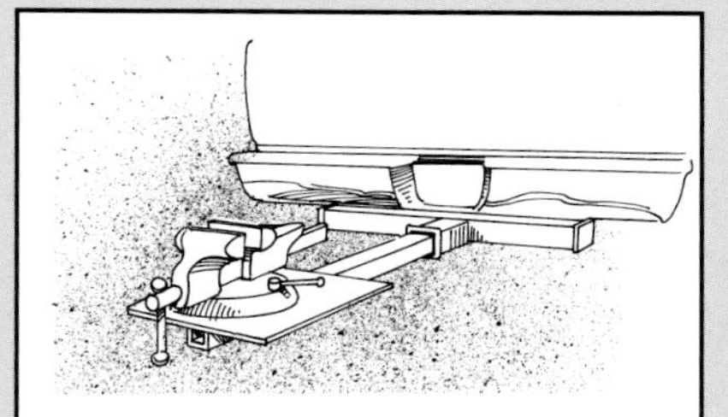

Quick-detach vise

This vise is mounted to a ⅜-inch-thick, 8x8-inch square plate, which is welded to heavy-duty 20-inch-long square tubing. The same tubing (like that in a standard pickup hitch) is welded under the welding table. The vise rotates in four positions. Remove the entire system by pulling one pin.

–H.F., Minnesota

Straight truck conversion

Our 1965 300-bushel, single-axle straight truck still had a good bed and hoist, but no motor or transmission. So we turned it into a pull-behind trailer.

After removing the cab, motor, and transmission, we bent the frame to a point and then added a hand-fabricated hitch made from scrap steel. The bed uses the tractor's hydraulics to raise and lower.

–N.M., Indiana

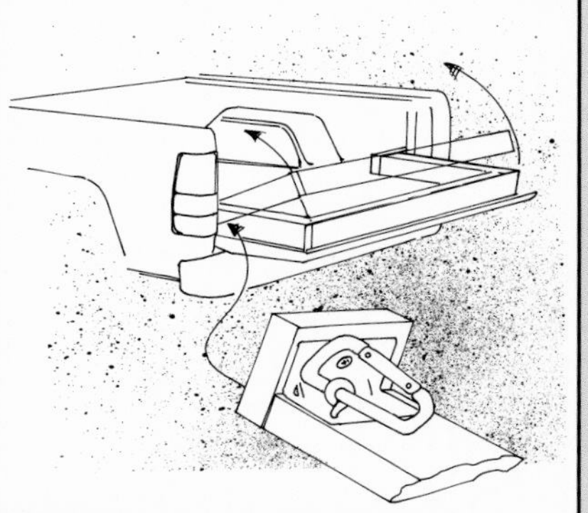

Make an extended bed for your pickup

My bed on my pickup was only 6 feet long. Often I needed to haul loose items like fence posts that were longer than 6 feet.

I built a retaining box from 2x4s that can fold up or down over the tailgate. It is locked in place by using snap hooks that attach to the tie-downs at the bottom of the box. Keeps items secure so they don't slide out the back when loaded.

–M.P., Ohio

Keep rodents out of vehicles

I keep my grain truck in the barn, and rats climb up onto the transmission. They chew holes through the rubber boot and crawl into the cab to eat the soybeans they bring in.

Fed up with replacing these boots, I inserted a sheet of aluminum under the floor mat and rubber boot. This shield isn't fastened down anywhere, so it's free to move in case of engine vibration.

–J.K., Ohio

Get extra strength without welding

Bolt bumper to receiver hitch with two 3½x5¾-inch pieces of ⅜-inch angle iron. Bolt through bumper's knockout holes, holes in the top and bottom of square receiver. Make a ½-inch hole for the pin.

–P.S., Oregon

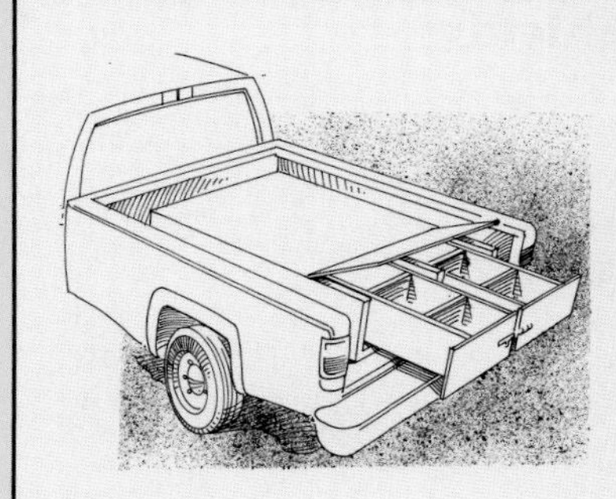

Rollout truck drawers

Each drawer in this toolbox is on high-quality roller bearings. They run on channel iron to pull out easily no matter how heavy the drawers may be. Dividers can be removed to accommodate any tool. I also added a 2¼-inch lip on both sides of the top to keep tools from falling off.

–H.F., Minnesota

Stepped-up storage

Systems for handling and drying grain need to be designed to grow

Grain storage systems may be made of concrete and steel, but they need to be flexible. That doesn't mean they move in the wind, but that they change with the times to stay in step with farm size and harvesting capacity.

South Dakota farmer Scott Campbell designed his system with that flexibility in mind.

"We had the chance to start from scratch because we moved to this farm in 1998 and grew our first corn in 1999," says Campbell. "That's a neat opportunity, but it does put pressure on you to plan for the future."

Photographs: Mike Hood, Rich Fee

South Dakota farmer Scott Campbell says open space is key, but there's more that goes into planning a grain-handling system for future expansion.

Drying capacity with a plan

Campbell's system includes three storage bins, a wet holding tank, an overhead loadout bin, and a continuous-flow grain dryer. A 5-inch, pneumatic-handling system delivers grain from the dryer to the storage bins, while augers serve the wet holding tank, dryer, and loadout.

"Our first decision was selecting a dryer," says Campbell. "We decided on a NECO Model 1680 because it was the least expensive way for us to get the ability to cool grain in the dryer without going to a batch mode and reducing capacity. We did this when we only had one storage bin. But now we use dryeration, dumping hot grain at 17.5% and pulling remaining moisture off while cooling in the bins."

Warren Odekirk, facilities planning and supply manager for Growmark in Bloomington, Illinois, says dryers should be sized to accommodate at least 25% growth, plus a safety factor for harder drying conditions.

"We start by recommending a dryer with at least enough capacity to dry a day's harvest in 18 hours. Then we subtract two or three hours for the safety factor," he says. "If you add acres or a bigger combine, you can run the dryer 24 hours per day."

Campbell says, "It's probably wise to pour cement and install the electrical capacity to accommodate your dream dryer from the outset."

Campbell feeds his dryer from a wet holding bin filled by a 12-inch auger that swings under truck hoppers. "This works and was economical, but it is one of the first improvements we plan to make. We want to put in two 500-bushel drive-over dump pits with an auger dumping to a short above-ground

Platforms on both sides of the loadout bin allow trucks to be filled without standing in dirt and dust.

The dryer control and pneumatic intake of Campbell's system face the prevailing wind to stay clean.

leg. This would let semitrailer trucks unload in a hurry and get back to the field," he says.

The pneumatic system that uses air to transport grain from Campbell's dryer to the storage bins was also chosen with expansion in mind. "It will be much cheaper to extend the tubing to reach additional bins than to add downspouts and cross augers from a grain leg," says Campbell.

At distances of 50 to 75 feet, the air system has a 1200-bushels-per-hour capacity. As the distance increases, that capacity will decline.

"We use two 10-horsepower motors to power the system," says Campbell. "A grain leg would have needed larger motors and required three-phase power, which would cost us $50,000. We also like it that maintenance work on the pneumatic system is on the ground, not 70 feet in the air."

Odekirk says pneumatics are especially suited for systems using continuous-flow dryers, where high capacity isn't a concern. "Their popularity is expanding rapidly," he says.

More fine-tuned features

Campbell raised his dryer to dump right into his pneumatic system, thus eliminating the need for a jump-auger to fill the system's air lock.

He has also wired automatic shutoffs, timers, and grain-level indicators into the system for safety and control. "We installed loops of electrical conduit at each of the electrical boxes with one leg of the power to the different motors running through it," he explains. "This lets us use an ammeter to sense the load on each motor."

Another highlight of Campbell's system is a 1,900-bushel overhead loadout bin. "A lot of truck drivers don't like to haul off the farm because of the delay," he says. "But here they can load faster than at most elevators. We can load in two to three minutes if the driver is up to it."

By Larry Reichenberger

By now it's clear that farmers have plenty of good ideas. Sometimes, however, their ideas are bigger, that is to say, they require more explanation and illustrations than space permits on the *All Around the Farm* page.

So we asked our readers to share some of those concepts in a program called *Bright Ideas*. Specifically, we asked for time- and money-saving equipment ideas.

And they responded – there were stacks of envelopes with photos, videotapes, letters, and drawings. We got ideas about sprayer makeovers; planter and air seeder modifications; truck and trailer transformations; and ways to custom-build field service vehicles and repair equipment.

They sent tried-and-true grain hauling, handling, and storage ideas; tillage tool makeover projects; tractor and combine modifications, and of course, shop tools, features, and designs.

Once again, farmers demonstrated that they are most interested in hearing from other farmers. The *Bright Ideas* stories consistently rated among the most read and useful stories in our reader surveys.

These are stories a little larger in scope and size, but still conceived, built, and put in practice by farmers who read *Successful Farming* magazine.

Weigh grain from the truck cab

The farther from home they get during corn harvest, the more hauling becomes a bottleneck for Sam, Roger, and Bob Ellis, Chrisman, Illinois. Based on the premise that every minute counts, the Ellises devised a simple way to cut the time it takes to weigh and unload trucks at their dryer.

Before, the truck driver had to get out of the cab and go into the scale house to weigh the load. Then, after dumping the load, the driver had to get out of the cab again to weigh the empty truck.

Now, the driver simply turns a switch from the cab to weigh the full load. After dumping, the driver backs onto the scales and reweighs the truck the same way.

Sam figures it saves at least four or five minutes every trip. "Plus, it saves a lot on the legs," he says.

While the grain is being dumped, the driver tests the moisture and enters it on the computer tape. ■

Chad Ellis weighs a load of grain without leaving the truck.

Photographs: Mike Holmberg

Darrin (left) and Chris Jensen converted this beer truck into a convenient chemical and seed trailer.

Roll out the barrels

A recycled beer truck is the basis for this lockable trailer

Darrin and Chris Jensen take a bit of ribbing when they park a beer truck at the end of one of their fields. But they say the advantages of their recycled beer truck for hauling chemicals and water more than compensate for the ribbing.

"This has greatly increased the efficiency of our water hauling since we can travel at greater speeds between fields or back home to fill," says Darrin, of Elmore, Minnesota. "There is no worry about loose objects flying off, and it trails better than a flat rack. And the doors lock so I don't have to bring all the expensive chemicals home every night."

The brothers originally went looking for a van trailer to haul chemicals and water, but found the beer truck. The low-slung trailer required a bit of remodeling to accommodate the two 1,650-gallon water tanks.

The low step-in gives the Jensens easy access to their mixing cone and chemical supply.

The first problem was removing a steel wall running from front to back in the center of the trailer. Since that wall also supported the roof, they ran several pipes from one side of the roof to the other for reinforcement.

Once they had the center wall out, Darrin and Chris ripped out the existing floor and had a local welder mount two steel beams to support a new floor. They covered the floor with 2×12s salvaged from an old corn crib. The total cost was about $4,000.

In addition to the main area, the trailer has smaller compartments on each side in the front and rear that can be used to store chemicals, ammonium sulfate, seed, and protective gear.

By Mike Holmberg
Farm Chemicals Editor

Charles Thor (below) converted this John Deere 3970 pull-type chopper (left) into a mounted unit for Ray Nietfield, Lake Henry, Minnesota.

Photographs: Rich Fee, Manufacturer

Semi self-propelled

Minnesota inventor converts used pull-type choppers into mounted ones

Two years ago, a heavy snow collapsed the roof on the machine shop Charles Thor had operated for 36 years in Hutchinson, Minnesota. (That snow may have been in revenge for all the tractor-mounted snowblowers Thor built in the 1960s and '70s.) Undaunted, Thor promptly had a new shop built on his farm at the edge of town.

"When you're 87 years old and you build a new shop, people wonder what's wrong with you," Thor says with a chuckle. "But I'm not done yet. I've got to have something to do."

Among other things, he has a few refinements in mind for the way he mounts pull-type choppers on Versatile and Ford New Holland bi-directional, four-wheel-drive tractors.

Several manufacturers build four-wheel-drive, self-propelled choppers. "But they get up over $275,000," says Thor. "This machine is not for the big outfits. It's for the average-size farmer who wants a self-propelled chopper." And, he adds, they can unhook the tractor in just a few minutes and use it for other things the rest of the year.

"Of all the stuff that I have made, that is the best piece of the whole bunch," he says. "You're carrying that chopper on those four big tires instead of pulling it. Boy, if you can't go with that, you'd better stay home."

In wet fields, it also helps that the wagon is being pulled closer to the tractor, says Thor.

Straight pipe

Nevertheless, it did take a few tries to work the bugs out. For example, on his first conversion, Thor ran the pipe up and then back. Unfortunately, it plugged, partly because a lot of the crop fed into the center of the fan and partly because of the curve in the pipe.

Now, he changes the way the crop feeds in and uses a straight pipe. "The centrifugal force keeps the silage to the outside of the fan and gets it moving," Thor explains. "And the straight pipe shoots it out without plugging."

Thor says it isn't especially difficult to mount a pull-type chopper on the front three-point hitch of a bi-directional tractor.

"You take the running gear and cut the axle off. You don't need that," he says. "You take the hitch off. You don't need that. You discard the long power-take-off (PTO) shaft. You don't need that.

"Then," he adds, "you have to make a unit that you can hook up with the three-point hitch on it. Nothing serious – just bracketing."

To finish up the job, Thor uses new parts that are readily available to modify the drive system. The drive on the machine shown utilizes an enclosed roller chain. Thor has also used a "power band" of five V-belts on some of his other conversions.

When he is done, the machine can be hooked and unhooked from the tractor cab, except for the PTO shaft.

Thor, who can be reached during the day at 320/587-2380, wants to put a three-row chopper on a late-model bi-directional tractor that has more horsepower than earlier models.

"I have got to make one of those," says Thor. "Then I am going to retire."

By Rich Fee
Crops and Soils Editor

Farmer field test confirms the benefits of calibration

Darrell Geisler is a firm believer in rebuilding his seed meters each winter. But he wondered if it paid to have those meters calibrated for accuracy.

So last spring the Bondurant, Iowa, farmer put the value of calibrating seed meters to the test for *Successful Farming* magazine.

To begin the effort, Geisler took his planter's finger pickup meters to neighbor Kevin Kimberley's shop to be rebuilt with new parts. However, only eight of the units were calibrated on the test stand. The other eight finger pickup meters were set according to the operator's manual.

The meters were split half and half on either side of his planter. Geisler then planted 20 acres with one hybrid in both sets of meters.

Local agronomist Brian Kolln of Heartland Cooperative later checked the plot and pegged the population range at 32,000 to 37,000 plants per acre. "While planting the (planter's) monitor was telling me the meters were working well. That was not the case," Geisler found. "As we learned after the corn emerged, we came to the conclusion that the corn kernels were passing the sensor so fast, it could not detect the extra kernels."

More doubles, triples

The resulting stand told the story. "What we've learned over the years was confirmed. If you don't run your units on a test stand, you will have a whole lot of skips, and even worse, multiples," Geisler says.

Yields, weighed by Heartland Cooperative, were surprisingly close. The calibrated meters produced corn that yielded 192 bushels. Nontested units went 194 bushels. "The only reason that happened is because of the extra population," Geisler contends. "Also, corn planted with nontested rows had 50% more lodging of the stalks, which definitely slowed down harvest."

Stands resulting from the field test confirmed that evenly spaced plants (7 inches apart) produced larger ears. They averaged 16 and 18 kernels around. All three ears in a test sample were 32 kernels long and produced 1,600 total kernels.

The stands produced by noncalibrated meters were uneven, often resulting in double and triple plant spacings. The resulting ears were 14 kernels around and ranged from 18 to 28 kernels long. Such ears produced a total of 1,344 kernels.

Accuracy pays dividends

The noncalibrated units also consumed more seed. "The 3,000 extra plants per acre cost $4.68 in seed. Sure, those stands produced 2 bushels more per acre. But I barely got my seed money back," he adds.

Also, Geisler believes multiple plants increase vegetation growth, drawing down soil nutrients. "Since you fertilize for X amount of yield, the extra plants cost you money," he says.

Besides confirming that calibrating meters pays dividends, Geisler also found that large round seeds seem to help minimize planting spacing problems.

By Dave Mowitz
Machinery Editor

Photographs: Mike Boyatt

Crafty caddy

This 20-foot coulter caddy features a bar that folds for road transport

Joe Vinton wanted to turn his 15-inch bean planter into a no-till rig. He needed to do something about the stalks since he didn't have any disks or row cleaners on the planter he'd built a couple years ago.

Adding a coulter caddy looked like the solution. "The 16-row planter is pretty heavy on the three-point hitch of the tractor, and I figured the caddy would solve that, too," Vinton says.

Vinton wanted a cart that was steerable so it would flex on terraces and keep the coulters in front of the planter units better than he could with a rigid bar. The caddy locks into a rigid position for road transport.

He started with a 20-foot rigid Great Plains coulter caddy and replaced the bar with a folding anhydrous bar he bought at a sale.

Vinton says he could have hinged the rigid bar that was on the caddy, but found the anhydrous bar (already hinged) available at a cheap price. It even had the same size tubes as the rigid caddy bar, so he didn't have to change any clamps.

Extending the hydraulics

He had to move the hydraulics for the air hopper of his planter to the back of the coulter cart. He also had to run auxiliary hydraulics to raise and fold the planter.

Vinton needs to add stops on the hydraulics to keep the caddy's tires on the ground, so it doesn't overload his planter gauge wheels.

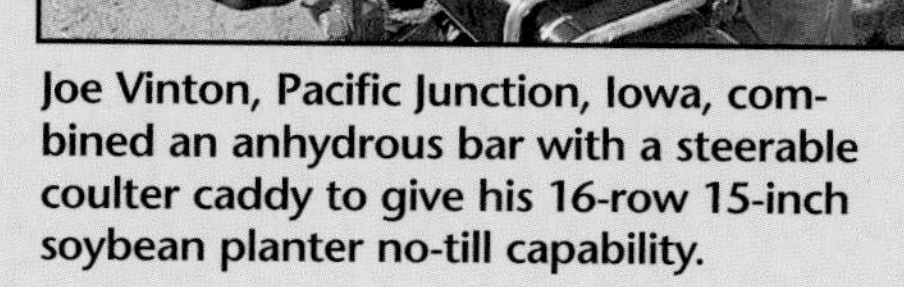

Joe Vinton, Pacific Junction, Iowa, combined an anhydrous bar with a steerable coulter caddy to give his 16-row 15-inch soybean planter no-till capability.

By Mike Holmberg
Farm Chemicals Editor

Narrow rows on a budget

Tim Walz switched from 38-inch rows to 20-inch rows by combining two planters and two corn heads

Tim Walz used row units from two four-row corn heads to build this seven-row corn head for 20-inch rows.

A neighbor uses the head on his John Deere 9500 combine.

Walz built one seven-row planter out of two four-row planters.

Photographs: Mike Hood

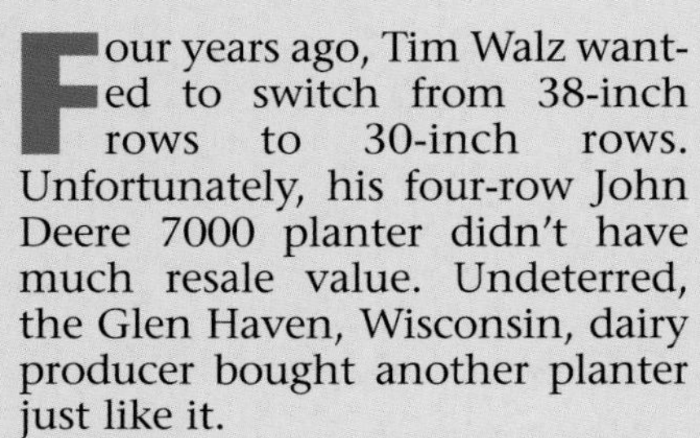

Four years ago, Tim Walz wanted to switch from 38-inch rows to 30-inch rows. Unfortunately, his four-row John Deere 7000 planter didn't have much resale value. Undeterred, the Glen Haven, Wisconsin, dairy producer bought another planter just like it.

Here's why: By combining those two planters, he built a seven-row 20-inch planter for less than he would have had to pay for a used six-row 30-inch planter.

Walz took the same approach to building a corn head. He combined two four-row John Deere 444 corn heads into one seven-row corn head.

Balanced budget

In addition to the one planter he had, Walz spent about $10,000 on the project. Most of it went for the second planter, two used corn heads, and seven new poly snouts for the corn head he built.

He also spent some time, of course, but the project was simpler than it sounds. Walz simply moved the two wheel assemblies to the outside edges of the planter. That freed up positions for two of the three row units he wanted to add to the original four-row planter. The third row unit was mounted in the center of the planter, "pretty much over the top of the transmission."

Heads up

Modifying the corn head was equally simple, but more time consuming. One of the heads he bought was set up for 38-inch rows, and the other one was set up for 36-inch rows. However, that series of head could also be set up for 40-inch rows.

When the row units are 40 inches apart, there's just enough room to squeeze another unit in between. "I just bolted them in and slid the six-sided shaft back through them," says Walz.

The only things that had to be altered to accommodate the 20-inch rows were the snouts. "You could probably cut the tin ware down, rebolt it, and get by," says Walz. "But I wanted to switch to poly snouts."

The seven poly snouts accounted for a little over 20% of what Walz spent on the project.

Walz hires a neighbor to harvest his corn. That person uses the seven-row 20-inch head Walz built on his John Deere 9500 combine.

Walz hires a different custom operator to harvest silage in 20-inch rows with a Kemper head. "Rows don't mean anything to them," he says. "They just come in and scissor it off." Walz says yields have increased with the 20-inch rows, and weed control has gotten better – and cheaper – due to the quicker canopy.

By Rich Fee
Crops and Soils Editor

Below: After purchasing the used van trailer, Jeff Fisher and son Erik set about modifying it for field use.
Below left: They added storage shelves for insecticides under the rear door as well as a pumping station under the middle of the bed. Left: They later salvaged a truck's hydraulic tail lift and added it to the rear of the trailer for loading seed corn and holding in-field transport.

Photographs: Mitch Kezar

Feed-and-seed center on wheels

Up to 100 acres worth of planting supplies go to the field

A $1,650 investment in a 40-foot-long semitrailer gave Jeff Fisher all the space he needed to take a day's worth of planting supplies to the field. "The trailer can hold up to 100 acres of seed and fertilizer, which is almost always enough to keep me supplied for a day," he says.

The Eaton, Indiana, operator puts a full complement of fertilizer in the form of 28% nitrogen and 10-34-0 through the planter. "I do some dry spread of potash in the fall," he explains. "Otherwise, all fertilizer goes on at planting."

To feed a fertilizer-hungry planter, Fisher gave the 40-foot-long trailer three 1,650-gallon tanks. Each tank holds 28% nitrogen. A fourth 1,650-gallon tank is storage for 10-34-0.

After the tanks were installed, Fisher was left with 8 feet of area at the back of the trailer, which offered room to hold a pallet of seed corn.

Customized features

Fisher rigged sight gauges, made from ¾-inch clear hose, for each tank and mounted them on the outside wall of the trailer.

"I can see tank capacity at a glance without having to crawl inside the trailer. I installed doors [obtained from a camper supply store] on the sides of the trailer to provide access to each of the tanks' shutoff valves," Fisher says. "I close those valves before transporting the trailer for safety sake."

All the tanks feed to a manifold system located under the trailer's bed. There are two 2-inch-diameter, 5-hp. pumps that Fisher uses to pump fertilizer to the planter or refill the tanks back at the farm.

As an added touch, he plumbed air from the truck's brake tank to the pump. This allows him to blow fertilizer out of supply lines after refilling the planter.

A 15-gallon sprayer unit (complete with electric pump, hose, and spray gun) is also mounted under the bed for planter or trailer cleanup in the field. On top of the trailer are spotlights for night work.

By Dave Mowitz
Machinery Editor

Strip-till boost

Deep-banded fertilizer adds bushels

Photographs: Josh Sears

Strip-till makes no-till corn work for Ashland, Illinois, farmer Greg Lepper. His 16-row bander injects dry fertilizer and NH_3 in bands 30 inches apart.

A toehold is a welcome find on a slippery slope. Recently, no-till corn yields in some areas have been on such a slope. Cool, wet springs and dry summers have challenged the practice.

But no-tillers are warming up to the practice of placing a band of nutrients in the root zone during fall strip-till operations. This gives the toehold that can stop no-till yields from sliding.

Strip tillage prepares a narrow width of the seedbed for next season's planting. The strip warms and dries faster in the spring, overcoming the slow growth that plagues traditional no-till. Adding a band of phosphorus and potassium to these strips improves fertilizer efficiency and encourages deeper root development.

Sizing up the benefits

Ashland, Illinois, farmer Greg Lepper put deep banding and strip tillage to the test last year. "We compared no-till corn planted on strips made the previous fall. Strips that received 200 pounds per acre of both 18-46-0 and 0-0-60, along with 160 pounds of nitrogen, were compared to strips where P and K were surface applied. We found that banding the P and K added 12 to 15 bushels per acre to our yields," says Lepper.

"It was a dry year, and we believe the yields increased because we got the nutrients deep in the soil where roots could use them for a longer period," he explains.

Lepper places the nutrient bands 8 inches deep with mole knives on his homebuilt 16-row strip-till unit. Dry P and K fertilizer is supplied from a Flexi-Coil air cart. A double-frame toolbar was built from 4×6-inch tubing. Three-point linkage between the cart and toolbar is connected to rephasing cylinders on a lift wheel on each wing and a lift-assist wheel on the rear.

Corn yield response

Bu/A: 120, 125, 130, 135, 140, 145, 150, 155, 160

N,P,K | N | N,P | N,P,K | N,K

Surface broadcast -------Strip-till band-------

Graph: Adding P and K to strip-till bands boosted yields over broadcasting or strip-tilling with just nitrogen on no-till corn.

The graph above details the results from a similar study by Lepper's local Lincoln Land FS dealer. On a no-till field that was rotated from soybeans, corn received 200 pounds per acre each of urea, 18-46-0, and 0-0-60. When those products were broadcast on the surface, corn yielded 134 bushels per acre. When only nitrogen was banded while strip-tilling, the yield was 143 bushels per acre.

P and K payoff

Adding P and/or K to the band with nitrogen further boosted yields. Banding only P yielded 150 bushels per acre; using only K netted 157 bushels. With both P and K in the band, the yield was 154 bushels.

Below: A homemade retractable hitch on the toolbar puts the controls for the hydraulics at Greg Lepper's fingertips.

The yield response Lepper is seeing differs only slightly from findings by Iowa State University agronomist Antonio Mallarino. "In no-till, we've found that strip-till alone may provide some yield benefit. But the largest and most consistent response is when potash is deep-banded in the strips. In long-term experiments we see an average yield increase to potash of about 5 bushels per acre," Mallarino says.

"Including phosphorus in the strip-till band seldom shows a response in our studies. However, there's no harm in including it, and it does make for an efficient application," he adds.

By Larry Reichenberger

Strip-till makeover

This used Kinze planter frame now carries fall strip-till row units

Eventually, planters get to the point they only have a little life left in them. Yet, the planter bar itself may have lots of life left in it. That's when the makeovers begin.

Brothers Steve and Terry Kutzley, Morenci, Michigan, bought a used Kinze planter and sold the planter row units to a salvage dealer. Then they purchased new strip-till row units from Progressive Farm Products, Hudson, Illinois (309/454-1564). They also bought new hydraulic lift cylinders for the frame.

Photographs: Andy Sacks

Steve Kutzley (right) and his brother, Terry, built this 12-row strip-till machine on a used Kinze planter frame. They like to pull it about 6 inches deep at 4 to 5 mph. That makes a mound about 6 inches tall.

New and used parts

By using a combination of new and used parts and doing the work themselves, the Kutzleys ended up with a field-ready 12-row strip-till machine. And it only cost about as much as a new six-row machine.

"A brand-new six-row machine was going to cost about $7,000," says Steve Kutzley. "When it was all said and done, we ended up with about $8,000 in the 12-row strip-till machine."

Each strip-till row unit has a pair of wavy coulters in front followed by a mole knife mounted on an anhydrous ammonia shank. A pair of 18-inch closing discs brings up the rear.

"We're always looking for a way to save time," says Kutzley. "Not having to till the soil is one way to get into the field quicker and get that crop planted. We had tried pure no-till, but the ground was just too cold and wet."

The Kutzleys plant with a John Deere planter equipped with two Rawson coulters per row.

There was another incentive, also. Converting from mulch tillage to strip-till qualified them for a cost-share program in Michigan. That program pays them $10 an acre for strip-till corn plus it cost-shared a storage building for herbicide and fertilizer.

By Rich Fee
Crops and Soils Editor

One advantage of the Kinze planter frame is its narrow transport width.

The strip-till units mount in the same location as the planter row units they replaced.

Strip-till goes mainstream

Once an obscure practice, strip-till is now growing rapidly

Strip-till corn has its roots in central Illinois and northwest Indiana. Paradoxically, much of the recent growth has been in the often wet northern Corn Belt and the usually dry plains where Kansas, Nebraska, and Colorado corner up.

Jim Kinsella, Lexington, Illinois, is widely credited with being the first Corn Belt farmer to practice strip-till. His first efforts in the fall of 1983 were tentative, but enlightening.

Kinsella modified an anhydrous ammonia (NH_3) applicator to better handle soybean residue and added some chain markers to guide it. That first machine had seven shanks. But because Kinsella had a four-row planter then, the rows of corn he planted in the spring of 1984 didn't always end up in the tilled strips. (The raised mounds that today characterize strip-till throughout much of the Corn Belt wouldn't come until later.)

"Every pass, unless I accidentally got my chain just right, had one, two, or three rows off the strips," he recalls. "But I learned a lot. There was a huge difference in early growth and probably a 20- to 30-bushel yield difference." Kinsella credits the yield difference to the fact that the spring of 1984 was wet.

Photographs: Andy Sacks, Rich Fee

Steve Kutzley (above) and his brother Terry started practicing fall strip-till two years ago. Because the soil is often cool and wet in the spring in their part of Michigan, no-till corn had been troublesome.

Modified no-till

Steve and Terry Kutzley of Morenci, Michigan, know a thing or two about cool, wet springs. "We wanted to no-till," says Steve. "But our ground is too heavy. No-till soybeans have been great, but we haven't had very good luck with no-till corn."

Strip-till looked like a way to make no-till work on their farm. (Some people like to debate the point, but many variations of strip-till count as no-till because the definition of no-till allows tilling a narrow strip to inject fertilizer. In terms of soil disturbance, strip-till is similar to the way many no-tillers inject anhydrous ammonia ahead of planting even though they don't plant into those tilled strips.)

Time savings

Two years ago, the Kutzleys added strip-till units to their used 12-row Kinze planter frame.

"We are always trying to find ways to save time," says Steve. "Not having to till the soil in the spring ahead of planting is one way of getting into the fields quicker and getting that crop planted in the small window of time we've had. It seems like that window gets smaller and smaller every spring."

Kinsella has been both a student and teacher of strip-till for these past 20 years and keeps his finger on the pulse of change. He says the interest in strip-till has increased a lot recently, and he estimates that acreage is now doubling each year.

"It's like the late 1980s and early 1990s when no-till drill sales were doubling every year," says Kinsella. "There has been real growth in the northern half of Indiana and over into Ohio where they have had three or four really wet years in a row. And there is another surge out West."

Kinsella says there are significant differences between wet areas and dry areas in the reasons for strip-till, the way it is practiced, and the equipment that is used.

"In Illinois, Indiana, and Iowa, it is replacing no-till," he says. "No-till fields just don't dry out some years.

"In the West, it is replacing tillage," he adds. "They're just clearing a strip, and they're doing it to conserve moisture. That's a big difference!"

Digging deeper

Another new approach in both wet areas and dry areas is to use some type of deep-till shank in an effort to alleviate soil compaction.

Most of the first commercial strip-till rigs that began appearing in the late 1980s consisted of a cutting coulter, an NH_3 shank, and oversized closing discs. Those closing discs would catch the soil as it exploded off the NH_3 shank and form a mound of soil and air.

The residue, which is lighter than the soil, would land on top of the mound and protect it over the winter. Since the early 1990s, many strip-tillers have used a mole knife in order to make a taller mound.

Parallel tracking systems

As more companies enter the strip-till market, there are more and more variations on the theme.

"Probably the biggest revolution now," says Kinsella, "is the advent of parallel tracking systems." He says these "decouple the planter from the strip-till machine."

That's significant because a lot of people who want to plant with 12-, 16-, or 24-row planters don't want to buy a tractor capable of pulling a similar sized strip-till machine. As guidance systems continue to drop in price, this trend is likely to accelerate.

"What a lot of farmers are doing is buying a 16-row planter and an eight-row applicator (strip-till machine)," says Kinsella. "Then, with a global positioning system, they're able to pull the planter and applicator with the same tractor."

By Rich Fee
Crops and Soils Editor

Jim Kinsella has strip-tilled for 20 years. He likes to leave a lot of residue on the mound in the fall to keep it from eroding over the winter. Trash wheels on his planter move the residue aside at planting.

Photographs: Larry Reichenberger; Courtesy of Carhartt

Fill drills in a flash

A little ingenuity changes a semitrailer from a grain hauler to a field tender that speeds seed and fertilizer refills

There are two basic rules of farm productivity that apply to drills and planters big and small. The first rule: Don't stop! Obviously, the size of seed and fertilizer boxes dictate that a pause is required for refilling. Thus the second rule: Don't stay stopped!

Minneapolis, Kansas, farmer Brent Kindall subscribes to these tenets and has built a seed and fertilizer tender that helps him comply with them. And when planting is finished, he turns the tender into a grain trailer.

"We get dual use out of the rig by using it to haul grain at harvest as well as seed and fertilizer at planting," says Kindall. "We really like this versatility and that makes it more affordable. It carries 350 bushels of seed and 11 tons of fertilizer at planting and then carries 738 bushels of grain at harvest."

A hopper bolts under the grain trailer to guide seed and fertilizer to individual hydraulic augers.

Kindall had the rig's 26-foot grain box specially built with a divided interior. Both sides slope to unloading gates. A divided hopper makes the transition to the pair of 5-inch hydraulic augers that move the material to his grain drill.

Supplies to last all day

"The augers were standard tailgate folding units, except that we had to shorten the tube below the fold and add to the tube above the fold to get the correct height," explains Kindall. "When we fold the augers down, one goes forward and the other backwards, and they fit into racks against the trailer."

Kindall's trailer carries enough seed and fertilizer to last all day – or even longer – with his 30-foot drill. He refills the trailer at night to prepare for the field the next day.

The augers are 15 feet long with 10-foot downspouts. Since they're mounted solid Kindall does have to move the truck to fill. "We fill with seed, then change power to the other auger and fill with fertilizer. Then we move the truck and repeat the process. It's a one-person operation," he says.

It also hauls grain

"It's a big improvement over the hopper box and single-axle truck we used before. We're not stopped as long, and we don't have to go after seed or fertilizer during the day. After planting, we unbolt the hopper, remove the augers, and we can haul grain," says Kindall.

Kindall installed a PTO pump on the truck to provide hydraulic power. The Freightliner, originally a Ryder moving van, was bought used without the box. The frame was shortened 10 feet to accommodate the grain box.

By Larry Reichenberger

Semi serious

With seed in a semi, the Jaquishes fill a planter in 10 minutes

Photographs: Grant Mangold

Thanks to the two elevators on the side of the semi he hauls seed in, Don Jaquish can fill this 12/23-row planter in about 10 minutes.

Don Jaquish figures "time in the winter is less valuable than time in the spring." That's why the Eau Claire, Wisconsin, farmer and his sons spent 200 hours and $1,000 or so to add elevators to the semitrailer in which they haul seed.

"The nice thing about it is we can back the 12-row planter up to the semi and fill all the boxes without moving the planter or the truck," says Jaquish.

The elevators started out in life as clean grain elevators on John Deere 95 combines. They have roller chains and rubber paddles that Jaquish explains are gentle on the seed.

Good as new

The Jaquishes completely rebuilt the elevators, which were donated by a neighbor, Ron Spehle. They also rebuilt the cross augers that take the seed from hoppers under the semi to the elevators. The shafts on the cross augers drive the elevator.

Each shaft is driven by an orbit motor. An 8-hp. engine drives a hydraulic pump that is actually "two pumps in one," Jaquish explains. "We can either run the front elevator or the back elevator individually, or we can run them both at the same time.

"We have run the elevators for two years and had absolutely no problems with them whatsoever," says Jaquish.

But they did make a major change the first rainy day. They mounted a magnetic solenoid valve on each elevator housing and ran a small tube down to a hole in the tube of each cross auger. That lets them dribble inoculant into the auger while they're filling the planter.

"Before we made that change, it only took 10 minutes to fill the planter. But it took another 15 minutes to put the inoculant on the seed," he explains.

Switches on the seed tube connected to the elevators control both the flow of seed and inoculant application.

By Rich Fee
Crops and Soils Editor

A solenoid valve controls the flow of inoculant to the cross auger leading to the elevator.

An 8-hp. engine provides plenty of power to run the hydraulic pump. That pump has two chambers. It can drive both pumps at once or one pump alone.

Don Jaquish's sons Jason (foreground) and Arthur team up to fill the planter.

Merle Doughty (right), Chillicothe, Missouri, had help from 11-year-old grandson Kellin when he converted a John Deere combine into this self-propelled sprayer that his son, Doug, uses.

Photographs: Mike Boyatt

Low-cost sprayer

For about $5,000, Merle Doughty built a self-propelled sprayer that meets his son's needs

Merle Doughty retired a while back, and he was looking for a way he could lend a hand with his son Doug's farming operation.

"I thought the thing that would help Doug the most would be to have a self-propelled sprayer ready to go to the field all the time," Merle Doughty says. "That way he wouldn't have to be mounting the tanks and boom on the tractor and then taking them off to do something else."

The price tags on the new spray rigs convinced Merle to try and build one instead. And since his grandson, Kellin, is mechanically inclined, Merle thought it would be great experience for Kellin to help build something like this.

Doughty started with a $500 John Deere Model 4400 combine and turned it into a self-propelled sprayer with a 42-foot boom that's just right for the terraced ground his son farms.

"The terraces are generally 120 feet apart, so I either had to go with a 60-foot boom or a 40-foot boom," Doughty says. "As uneven as the terraced ground is, I thought a 40-foot boom would work better. We made the boom 42 feet wide so we don't miss any spots."

The trickiest part of the sprayer project was tearing down the combine while trying to keep the frame and drivetrain all together. "When I got everything off, I had to start putting steel back on it to hold the cab and motor up," Doughty says.

"I also reinforced the frame with 2×6-inch tubular steel. By the time I got down to the basics, there was just a 4-inch channel iron connecting the front and back wheels. I knew that wasn't nearly heavy enough, so I put the tubular steel on top of that frame for extra support," he explains.

Support needed on the frame

"I had to put quite a bit of cross bracing underneath where the cylinder and separating mechanism had been, because when we took all that out, there wasn't enough steel to keep the machine rigid enough," he adds.

Doughty wanted the boom on the front of the sprayer for better visibility. "With the self-propelled unit, I could put the spray boom out front where it's visible at all times," Doughty says.

Doughty figured he could reuse the boom they had been using. It had been fitted with mouting brackets so it could either be mounted on the three-point hitch of the tractor or on the back of their pull-type sprayer.

To mount the boom on the new sprayer, Doughty added 6-foot lift arms that pivot from the yolks where the corn head had pivoted before. That put the boom about 4 feet in front of the cab for good visibility. He used the header cylinders from the combine to raise and lower the boom.

Stabilizers needed for boom

Doughty originally made the lift arms of 4-inch channel iron, but they found that the boom was swaying too much. To stabilize it, he added tubular steel reinforcements (see photos at right) between the lift arms.

To get the boom to fold, Doughty connected cables from the boom to the hydraulic reel lift from the combine. The cables fold the boom in from the cab. "My son says that's the best thing I put on it," Doughty says. "With our old sprayer, Doug always had to get out to fold up the booms. He says this is a real time-saver."

Doughty salvaged a 600-gallon tank from an old pull-type sprayer they had, and they mounted it at the back of the frame.

They considered putting a 1,500-gallon tank on the sprayer but couldn't find one with a deep enough sump. They also mounted a 30-gallon rinse tank under the engine compartment.

Doughty replaced the combine tires with tall, 12-inch tires to reduce damage to crops. His son normally sprays about 5 mph and can travel on the road at 20 mph.

The cab includes a MicroTrak spray controller, air conditioning, and a hydraulic suspension seat.

"My grandson is quite proud of what we did, my son likes the machine, and it didn't cost us an arm and a leg," Doughty says.

By Mike Holmberg
Farm Chemicals Editor

Top: The boom lift arms connect to the yolks where the combine header had connected. Doughty added tubular steel reinforcements to stabilize the boom. Bottom: Before he mounted the 600-gallon spray tank, Doughty added the 2×6-inch tubular steel to strengthen the combine's frame.

Photographs: Rich Fee

Left: Sam Ellis and his sons, Roger (blue shirt) and Bob (not shown), mounted an auger cart on an old combine chassis.
Above, top: This is the view from the cab of the self-propelled cart.
Above, bottom: There's ample clearance between the auger and trailer.

Cart with a view

This self-propelled auger cart makes it easier to load semitrailers in the field

From the cab of a tractor hooked to a grain cart, you can't always tell whether you're topping off the load on a semitrailer or shooting grain off the other side.

Sam, Bob, and Roger Ellis, Chrisman, Illinois, took care of that problem by mounting a new 750-bushel Brent auger cart on a modified International 1480 combine chassis.

"It has made loading trucks a lot easier and a lot faster," says Sam Ellis. "I can pull up exactly where I need to be. I don't waste time backing up or pulling ahead."

There's another benefit: The self-propelled cart keeps the Ellises from having to log an extra 200 to 300 hours per year on their $100,000 tractor.

The Ellises considered several makes of combines before deciding on the 20-year-old International 1480. "We chose that model because of the hydrostatic drive," Sam Ellis explains.

The 1480 combine is all direct drive. Consequently, there aren't any belts to deal with. There are only hydraulic lines, which are simple to hook up.

Figuring the cost

They bought the combine for $7,000 at an auction, then sold about $3,000 worth of unneeded parts. But, they also bought some parts, most notably heavier final drives out of a newer Case IH 1680 combine and a two-speed hydrostat that was original equipment on some 1480s.

The hydrostat works in unison with the three-speed transmission to provide faster speeds in the field and on the road. "In the field, when I'm running in second gear, I can go about 2 miles per hour faster than I could before," says Ellis. "That's loaded. Empty, I can run in third gear about anywhere if I need to catch up with the combine."

The heavier-duty final drives and the hydrostat were both just bolt-on changes.

Most modifications were confined to the combine chassis rather than to the auger cart. Every one of the modifications to the cart is easily reversible.

A section of tongue was unbolted and removed, and the wheels were switched from the cart axle to the combine axle.

The driveshaft had to be cut, "but we had a coupling made so we can put it right back together," says Ellis. All the hydraulic lines have quick couplers.

Ellis estimates they could have the auger cart back ready to pull with a tractor in about half a day, if

This view from the combine cab shows how the engine on the self-propelled grain cart nestles underneath the slope of the grain box. By turning that engine crossways, the whole rig is more compact.

Sam Ellis says it didn't take as long to convert the International Model 1480 combine into a grain cart as he thought it might. The hydrostatic drive made it easier to move things around.

need be.

The original combine engine was used, but it was turned crossways and mounted under the front slope of the grain box. That made for a more compact unit.

Fitting the pieces together

The unloading auger is driven by a belt that was originally used to turn the rotor on the International 1480 combine.

"We didn't have to modify anything to do that," says Ellis. "We just had to tilt the engine slightly to line up with the shaft that drives the auger." An electric clutch controls it.

The Ellises mounted the cab just as high as they dared, for better visibility when loading semitrailers.

All that effort does pay off during the long days at harvest. "Running the cart is the easiest job now," adds Ellis.

By Rich Fee
Crops and Soils Editor

Mini down-corn reel protects the center

You'd need a computer, or at least a thick record book, to keep track of all the equipment modifications the Ellises have made over the years.

Those modifications range from the simple to the sublime. But even the simple modifications can make long days go a little more smoothly and seem a little bit shorter.

Using their heads

Troubled by stalks building up at the center of the corn head on their John Deere 9600 combine, the Ellises spent a half day in the shop building the mini down-corn reel shown in the top photo. It's positioned above the four center rows on the 12-row head.

"Before, we'd have to back up, shut the combine down, and get out there and remove the stalks," explains Sam Ellis. "The reel is short and up close, so it just pushes the trash into the auger."

Top: This mini down-corn reel keeps stalks from building up at the center of the corn head.
Bottom: This is the view from inside the combine cab.

Photographs : Mike Hood

Richard Aufenkamp uses a leaf blower to blast the debris off his combine every day to keep from spreading weed seeds. He doesn't get inside the machine, but gets all the other places where the weed seeds gather.

Combine cleanout

A blow-dried combine doesn't spread weed seeds

A 200-mile-per-hour air blast from a lightweight leaf blower is just the ticket for cleaning shattercane seeds and other debris off a combine, says Richard Aufenkamp, Nebraska City, Nebraska. By cleaning his equipment, he avoids the spread of problem weeds.

"Every time we finish harvest on a particular farm, we blow the combine off just as much as we can to avoid moving weed seed," Aufenkamp says.

He has shattercane on two of the farms he operates but not the others and doesn't want to spread the seeds. "All the rest of the ground we have is relatively clean; we didn't want to carry the shattercane seed with us," he says.

Shattercane seed tends to collect on the combine, he says. "We started using the leaf blower about five years ago, and I think that has really helped. We're getting the shattercane under control."

Other weeds aren't such a big concern because all his fields have waterhemp, sunflower, and other normal weeds. Shattercane is more of a challenge, he says, because there are postemergence herbicides to use on it but no really good residual herbicides to deal with it. "If you're going to beat shattercane, you've got to fight it by not spreading it."

Aufenkamp prefers a leaf blower to an air compressor for cleaning the combine. "We're strung out, and this leaf blower makes it so convenient. The one we have puts out 200-mph air. It's a lot faster because it has quite a bit more volume than an air compressor."

A half hour of fire prevention

It takes about a half hour to blow the machine off, Aufenkamp says, and they try to do it every night. "I don't know if that's a lot of time or not if you consider the safety factor involved. I see some farmers let their combines build up with debris. It's no wonder they catch on fire. I'm scared to death of a fire. I guess I'd rather spend the 30 minutes it takes to clean it.

"I've been farming since 1970, and I've never had an inkling of a fire in a combine. If you blow it off every day, it doesn't take near as long to do the job," he says.

The toughest places to get at for cleaning are under the cab and the bin. But the weed seeds are not as likely to accumulate there anyway, he says.

"The leaf blower I have has kind of a small end on it, which makes it easier," Aufenkamp says. "You can get the sides of the machine, the axles, and the top of the frame relatively easily. But under the bin and behind the cab are a little harder to get at."

Aufenkamp is nearly as meticulous with his tillage and planting equipment. "In the spring when we're planting, we'll clean our equipment before we go from one farm to the next so we don't carry weed seeds in a dirt clod. We don't wash it down, but we do clean off all the loose dirt before we go from one farm to the next."

By Mike Holmberg
Farm Chemicals Editor

Photographs: Ron VanZee

Tony Roorda and father Darrell (on the right in the picture below) spent about 1½ winter months in their shop fashioning a self-propelled loader from a junked-out combine and an existing bucket loader.

Harvester now hauls the harvest

A salvaged combine now works as a loader

After replacing the clutch in his Oliver Model 1750 "at the rate of one clutch every year" and getting stuck in muddy lots because of the tractor's small tires, Tony Roorda was in the market for a new loader.

Roorda found a replacement in the form of an IHC Model 815 combine. "I went looking for an International because they had a foot-inch pedal," the Ypsilanti, North Dakota, farmer explains. "I thought this would be a good feature since your first reaction in a crisis is to push in the clutch."

The $1,800 harvester came home with Roorda and then was promptly stripped to its frame.

Next, he added a frame of 6-inch channel iron to the combine's chassis at the same distance "from the ground up to the new frame that matched the height my older loader was mounted on the Oliver," he says.

He positioned the combine's cab to the channel-iron frame. Then came the engine and fuel tank.

"Well, I needed weight over the back wheels, so that was where the engine went. The fuel tank went between the cab and engine," he says.

The conversion was finished by hooking up electrical, fuel, and hydraulic lines. "I used the three-spool valve off the loader and mounted it in the cab," he adds. "Originally I used the header height spool and hooked it up to the three-valve spool in the cab."

Added a high-output pump

The combine-turned-loader immediately went to work. After 1,000 hours of proving its worth, Roorda felt it was time to invest in an auxiliary hydraulic pump with its own reservoir to run the loader. "The original hydraulic pump on the combine just wasn't big enough," he explains.

While he was making improvements, Roorda decided to add a PTO to the loader. "To do this, we drive the PTO off of the separator drive pulley because it has a clutch to turn on and off," he says. "We run a belt off of that to a shaft to a chain down to the old drive of the unloader auger bevel gears from the combine (to switch operating rotation) to a shaft out the back of the PTO hookup."

$2,000 price tag

At an investment of around $2,000, not including the new high-output hydraulic pump, Roorda is in love with the loader.

"The visibility alone is a great feature. And no more broken front axles from the weight of the loader," he says. "We hay a lot of ditches, and retrieving bales from a ditch is great because the center of gravity is low and weight is distributed evenly."

Getting stuck in mud is a thing of the past, too. "The frame can be dragging in mud, and it still pulls through," he says. "It's a fun unit to run!"

By Dave Mowitz
Machinery Editor

Photographs: Andy Sacks

Tom (above) and Doug Burrer took a no-nonsense approach to equipping their shop. A salvaged hydraulic lift works with floor cherry pickers to substitute for more expensive overhead hoists.

Long on practicality

The Burrers build rather than buy their shop needs

Every winter the Burrer brothers burrow into their shop and meticulously go through all their machinery to rebuild, remodel, and revive the fleet.

That dedication to maintenance explains how the Elyria, Ohio, farmers can operate with equipment that is often 20 to 25 years old. "We take good care of it," Tom Burrer explains, "and it keeps operating beyond traditional trade-in life."

Every inch put to work

Tom and brother Doug credit their shop for making such maintenance diligence possible. That structure is the epitome of practicality. Every square inch of the shop is put to use whether in service bays, supply and tool storage, or a fabrication center.

Their approach to creating the shop? "We build it ourselves or salvage what we can before buying shop accessories like storage areas," Tom says. "Yet the shop has all the tools, equipment, and adequate room to perform any and all – I mean all and not just some – maintenance jobs."

For example, the Burrers didn't invest in an overhead swinging or trolley hoist. Instead, they purchased a salvaged hydraulic automotive floor hoist.

"We have used it to lift the front of semitrucks as well as entire cars or pickups," Tom says. "A cherry picker hoists everything else. Sure, an overhead hoist would be great, but we couldn't justify the cost."

Such is also the case with wanting a larger structure. "Really, don't you always want a larger shop?" Tom asks.

Yet the 48×62-foot structure can house up to three pieces of machinery at a time for servicing. That's possible because of the way the brothers allocate floor space.

"We created separate work areas that don't interfere with each other to accomplish a variety of maintenance and service projects," Tom says.

Three bays available

Two major service bays are located behind the shop's two overhead doors. A third bay is carved out around the hydraulic floor lift.

"We can be changing the oil on a pickup on the hoist, for example," Tom explains, "while doing an engine or transmission overhaul on a tractor in another bay. And dirty jobs that need washing are positioned in the bay at the end of the building and near our steam cleaner."

The Burrers prefer to wash their machinery outdoors. But an indoor wash bay, equipped with a floor drain, is a necessity in the winter. Also, this same location is used when any painting needs to be done.

This division of work space

allows the Burrers to call in a professional mechanic to tackle engine overhauls when necessary.

"This certainly cuts down on the cost of an overhaul," Tom explains. "We can do one here for little more than the cost of the overhaul kit ($2,500 to $3,000), as opposed to having a job done at a mechanic's shop, which can run over $8,000."

The brothers also trimmed their construction expenses when the shop was erected 10 years ago. They decided to build on the foundation of a previous building that had burned down.

"There was no labor and not much help from any contractors," Doug says. "We put all the nuts and bolts together for everything, including electrical, plumbing, ventilation, lighting, hydraulics, loft storage, drainage, concrete, and welding."

By Dave Mowitz
Machinery Editor

A low-cost battery charger wired to 15 leads provides a trickle charge for batteries to keep them in condition while they're in storage. The batteries sit on wood shelves to prevent discharge, a common occurrence when batteries are stored on concrete.

A centrifugal fan, salvaged from a forced-air furnace, readily draws in welding and grinding smoke then propels it down a 12-inch-diameter duct to the outdoors.

Even drain tile serves a second purpose by providing a home for long-handle tools. Pry bars, tire irons, and sledge hammers are organized in the tile containers for ready use.

Used oil is captured and held for proper disposal in homemade containers. Walls above and below the storage loft are lined with wood shelves and bins to keep parts in order and easily accessible.

A wide assortment of funnels, filter wrenches, and coolant equipment drips excess fluid into a trough made from a length of roof gutter. It resides next to jugs of specialized oil in the Burrers' lubrication center.

The loft serves as home for 440 gallons of lubricants in the form of engine and hydraulic oil. Two sets of drums are dedicated to a specific lubricant, which is fed to a central location on the ground floor to be tapped for use.

Photographs: Ron Van Zee

The handrail on Roland Wedel's staircase folds flat across the steps when the staircase is retracted for storage.

Retracting loft staircase

It rides on rollers so it can be pushed out of the way when not in use

Adding a loft to a shop is a great way to create a wealth of storage space. But to access a loft you've got to have steps, and oftentimes they stick out like sore thumbs and block floor traffic.

The latter situation was certainly the case with Roland Wedel of Willmar, Minnesota. He found that the staircase to access his loft would end up partially blocking the shop's main entrance.

Not to be deterred, Wedel did some creative engineering and fabricated a staircase that readily folds up and out of the way when not in use.

His innovation – built of steel tubing, wire panel, and strap – is mounted to the loft deck via 1½-inch steel rollers that feature bearings. These rollers ride inside ¼×1½-inch steel straps that act as guides when the staircase is retracted or extended.

Dolly wheels at the bottom of the steps make it a breeze to push the staircase into an upright position and up out of the way of his big equipment.

"The handrailing folds against the stairway in the upright position by releasing a latch," Wedel adds. "When using the stairs, the rail must be in the vertical position. Of course, this is another safety feature."

By Dave Mowitz
Machinery Editor

Roland Wedel's retractable staircase rides on dolly wheels. "In the upright position the wheels retract, so the stairs rest on the floor," he says.

Photographs: Ron Van Zee

Laurel Pistorius's rolling tool cabinet features electrical power and work space, as well as a storage cabinet and drawers and a labeled display board to keep hand tools easily accessible.

A rolling tool cart lets Eric Larson move tools and supplies to the repair site in his family's farm shop. A dog food dispenser on the end of the cart carries floor-sweeping compound.

Tools on the roll

Few things add more to a farm shop than good tools and a good place to keep them

Hanging shop tools on the wall or storing them in a workbench drawer assures you'll always know where they are. But it doesn't assure they'll always be where you need them.

By putting that tool storage area on wheels, some farmers have found a way to keep their tools on hand as well as at hand.

Portable tool cabinet

Laurel Pistorius likes his portable cabinet because it keeps his tools in view rather than buried in the bottom of a drawer.

"We've got hooks installed in the back of the cabinet, and each hook is labeled, so my grandkids can learn the names and sizes," says the Browns Valley, Minnesota, farmer.

Pistorius covered the top of the cabinet with metal and added a foldout shelf on the side for added work space. Two drawers hold sockets, while the cabinet underneath holds towels, shop rags, and various power tools.

"We also have wired the cart for electric power," says Pistorius. "We run an extension cord to the back of the cart, and the wiring runs to an outlet on the side. This lets us plug in lights and other power equipment while running only a single extension cord from the shop wall."

The tool cabinet is 4x2 feet deep. A display board on the back wall provides the area for hanging tools.

Rolling tool cart

The 80x80-foot shop on the Larson farm near Fullerton, North Dakota, is probably never bigger than when the Larsons have to walk across it dozens of times to gather the right tools for tackling a big project.

However, Eric Larson made that big shop small by building a cart that holds nearly every tool imaginable, as well as a selection of other frequently used supplies.

"We just wheel the cart right up to the project where we're working, instead of constantly going after tools," says Larson.

He built the cart using square tubing, caster wheels, and two pieces of 4x8-foot plywood.

Hooks on one side hold a complete set of hand tools, all painted orange to separate them from other tool sets.

The opposite side of the cart holds various supplies, including tape, wiring, clamps, and a paper towel holder. A cutout section provides a handy place to carry an assortment of boxes holding other commonly used supplies such as electric connectors.

"Two things we've learned are to leave room to add tools and keep heavy stuff – like bolt bins – off the cart," Larson says.

By Larry Reichenberger

Photograph: Le Spearman

A complete workbench of tools, electrical outlets, and lights ride out to work in the shop or in the yard on a tool cart that Myrl Waggoner fashioned from sheet and tube steel.

Wheels to work

Build this mobile workbench with common steel and a welder

Tired of walking around his shop gathering up tools required for a project, Myrl Waggoner created a workbench that travels to work. "That way, wherever I work I just take everything with me," the Oklahoma City, Oklahoma, farmer explains.

His mobile workbench is fashioned from common steel. "The wheels are the most expensive part on the cart," he says. "I got 6-inch rubber tires that will run over gravel so I can take the cart outside of my shop."

To build the cart Waggoner used a steel bender to form 12-gauge sheets of cold rolled steel into shelving and shelf supports.

Top and bottom shelves are 4 feet long and 2½ feet wide, while the cart stands 5 feet tall. Supports were made from 4-inch cold rolled steel tubing.

The top shelf of 12-gauge steel has a vertical panel welded to the back of the shelf. Waggoner utilized the back panel by making hooks out of steel rod that he welded on the front side of the panel. The hooks hold a complete set of filter wrenches.

Wrenches at the ready

Hung on the backside of the upright panel is a full set of metric wrenches. Attached rods that extend above this section hold socket heads.

Waggoner welded another metal frame with slots above the back panel to give him convenient screwdriver storage.

Waggoner made sure that he had access to electrical power at the cart by adding 110-volt electrical outlets, a troubleshooting light, and a halogen work light to his shop cart. He used a 110-foot extension cord, cut the plug end 40 feet long, and wired the cut end of the cord into an electrical connection box attached to the cart. This left him with a 70-foot-long female plug, which he connects to the 40-foot cord and an electrical outlet.

"Every time I needed an extension cord my kids would always have them. So I hard-wired the cord in so they couldn't borrow it," Waggoner says with a chuckle.

A vertical arm support made from 2-inch tubing holds both trouble and halogen lights. A retractable reel allows Waggoner to reel the trouble light out to where he needs it.

"I have a halogen light that I can use whenever I am working on a project," says Waggoner. "I can turn my workbench toward that project and shine the halogen light for full illumination."

Good neighbors

The bench also features a 6-inch vise and a bottom shelf of expanded steel, which allows dirt and debris to fall away from any tools and equipment.

Since Waggoner built his first cart in 1997, he has made five carts for neighbors. "I told my neighbors that I made them something that no other farmer could borrow," explains Waggoner.

By Le Spearman

Musings from the mud porch

Pack rats never worry, except over which way to use all that stuff

I have seen a lot of weird behavior in my life, people doing things I couldn't imagine in my wildest dreams. But you could have hit me across the head with a 6-inch fence post the other day.

I saw someone up in town (not to mention names), but she threw away a perfectly good 3-pound coffee can – complete with the plastic lid! Just threw it into the garbage without so much as a word of explanation.

Linda marked her calendar the day two unthinkables happened: I found my own billfold, and I threw something away.

OK, so I'm a child of the Great Depression. I can't leave food on a plate without thinking of the starving children in China my mother was always talking about. Our garbage man should give us half price since we generate maybe 1 pound of garbage a week.

Reuse, adapt, reduce

We may generate the average American volume of garbage, but we don't throw away the average American volume of garbage. And when we do throw something away, you can be sure the item has been so thoroughly worn out, reused, adapted, used again, converted, and reduced, that the only thing that makes sense for it is a well-earned rest in the dump.

Linda was pretty well trained by the time she came my way. Her folks are farm-raised Nebraska Czechs. You know, the kind that make sausage from everything on a critter but the feathers. And now that I think of it, the brush they use to put butter on kolaches is made of the feathers!

Roger Welsch

Anyway, Linda and I agree on this issue, even if we don't agree on music, tractor colors, or whether Cindy Crawford is cuter than Brad Pitt. She doesn't even bother to ask when I ask her to save eggshells for a couple weeks (perfect soil builder for some houseplants), to pile up milk jugs under the stairs (collectors for maple sap taps), or to pick up the cobs from the corn the deer and squirrels are eating at the end of the fence (for smoking hams).

Virtually no waste

A friend in town asked me to save my old overalls so she can make skirts out of them. As much as I'd have liked to help her, by the time overalls leave this place, they have been worn to the point of indecency. (As determined by Linda, not me. I would have guessed they still had 10 or 20 wearings in them.)

They're then reduced to 6-inch squares for cleaning engine parts in the shop, and finally, soaked in waste oil to serve as fire starters in the shop woodstove. For me, old overalls translates into ashes.

I don't recall ever throwing away one of those CD disks the on-line companies are always sending me in the mail, or a sample credit card. They're useful for mixing epoxy in the shop. The dogs love the cobs from sweet corn, and that way residue butter doesn't go to waste. My dentist buddy sends me his old dental tools for cleaning on engines.

Last night the deer and bunnies outside the yard got our artichoke leftovers. I fill the cat's bed with those white plastic peanuts fragile stuff comes packed in.

Fashion statements

But I think I'm most proud of something I invented myself. See, I wear cutoff sweatpants for pajamas, which leaves me a surplus of cutoff leg bottoms.

Now, I could use all these leg bottoms as rags, I suppose, but I have ingeniously figured out that I can first wear the warm tubes as make-do stocking caps. Sure, my hair sticks out the top, but the caps are very warm, and I think they look just fine when I go into town for the mail. Maybe even kind of sporty!

Hey, Linda, don't you think it looks kind of like one of those fancy ski hats when I wear a cutoff sweatpants leg on my head? Linda? Linda?

Hmm, now that I think of it, what happened to all those sweatpants legs I was storing out on the mud porch? I don't seem to find them. Linda? Do you know what happened to those sweatpants leg stocking caps of mine? Linda?

Roger Welsch, author, humorist and former correspondent for CBS Sunday Morning, *lives with wife Linda, daughter Antonia, a packof dogs, and a fleet of tractors in the Loup River Valley near Dannebrog, Nebraska.*

Photograph: Chester Peterson Jr.

All around their farms – and maybe yours?

These two longtime contributors have also invented equipment that may be in use on your farm today

Kent Woodford

Kent Woodford hates to make mistakes. But what he really hates is making the same mistake twice. It's this aversion to making mistakes that drives him, in large part, to invent things. That, and a penchant for saving time.

"With most of my inventions, what I'm really trying to do is save myself time or prevent myself from making the same mistake again. I attribute that to watching my father, Delbert Woodford, while I was growing up on the farm."

"When you're trying to solve a specific problem or make something easier or less time-consuming, it really becomes more a necessity to invent," he adds.

Woodford not only watched his dad, he was a quick learner. No fewer than six of his inventions have appeared in *Successful Farming* magazine's *All Around the Farm*. That may be a record.

He grew up on a farm, but by the time he and his three siblings came to adulthood, the home place wasn't able to sustain the entire family. So Woodford set off to find his own place in the world. He took some agronomy courses, eventually became a Certified Crop Adviser, and for the past 30 years has been involved in the seed, fertilizer, pesticide, and crop input industry. He also maintains a 20-head heifer operation near Macomb, Illinois.

Yetter seed chute

So what does Woodford consider his best invention?

"I guess my seed chute," he says. "That was the innovation I most enjoyed working on."

The chute is currently marketed by Yetter Manufacturing Company of Colchester, Illinois. *Successful Farming* magazine noted that besides easing the job of unloading, the Seed Chute targets the path of seed emerging from center dump seed boxes. The lightweight, enclosed metal channel deflects seed from the center seed gate opening of the box. The chute takes just seconds to add or remove without the need for wrenches. In addition, the installer doesn't have to stand under the box to fasten the chute. Installation can be made from the side of the box for safety's sake.

Backbone of America

Even though Woodford's not actively farming, he's glad he can make a contribution to agriculture by way of his inventions.

"I enjoy being associated with farming and believe it's the backbone of America," he says. "I just want to be involved in agriculture in any way I can."

Kent Woodford's effective time- and labor-saving ideas appear regularly in *All Around the Farm*. When he's not sending in submissions, he's working as a certified crop adviser.

Harold Fratzke

When Harold Fratzke isn't busy farming 400 acres with his son, you might find him tinkering in his shop. Nothing out of the ordinary there, except that Fratzke's definition of tinkering involves the invention of the Roto Chopper, Hydra Covers, Hydra Levers, and nearly 20 other pieces of equipment, devices, and innovations that have been marketed at one time or another.

Not bad for a "tinkerer." So why not call him an inventor? Because that isn't the way Fratzke thinks of himself. "I've always been a tinkerer," he says. "I'm just a regular farmer. Nothing fancy. No expensive fabrication tools."

Harold Fratzke farms near Cottonwood, Minnesota. But if he weren't farming, "I'd be in my shop working on new things," he says. (The rolling workbench is another of Fratzke's original designs.)

Hall of famer

OK then, so what is Fratzke's secret for coming up with ideas for products other people need? He'll tell you it's because he's a farmer. "The thing about farming is, it forces you to look for solutions to problems," Fratzke says. "That, and all the time spent daydreaming in a tractor, make a great environment for inspiration. But then, farmers are some of the most inventive folks around."

Inventive to the point that he's in the Minnesota Inventors Hall of Fame, along with Richard Drew of 3M, the father of Scotch tape and Post-It notes and Alexander P. Anderson, the inventor of puffed wheat.

Most of Fratzke's inventions are practical solutions to everyday problems. Occasionally, they're so practical, a common response is "why didn't somebody think of that before?" And this has led to some manufacturers copying several of his inventions, which doesn't really bother Fratzke.

"A patent takes a lot of time and money to secure," he says. "And even more time and money to defend. I spend my energies working with a manufacturer to get the idea on the market." He credits much of his success to his ability to convince manufacturers that an idea can be sold. "A manufacturer doesn't always see the potential in an invention," Fratzke explains. "So I argue hard. If that doesn't work, I've been known to go out and sell the idea myself just to prove it's marketable."

K & M Hydra Covers

Fratzke puts Hydra Covers at the top of the list of his most innovative ideas.

"Every farmer's faced the frustration of dirty hydraulic ports on tractors," Fratzke points out. "Hydra Covers are spring-loaded covers that enclose the ports, preventing contamination."

The invention proved to be so practical and popular that they're standard equipment on most tractors built today.

This is not to say that everything Fratzke dreams up works out for him the first time. But that doesn't mean he loses sleep over it.

"Sometimes, when I've invented something and have a problem with it, I'll go to bed and wake up with the answer."

Which makes Fratzke an inventor 24 hours a day.

10 TIPS
from one frugal family

Secrets from a Nebraska woman who ekes out a middle-class lifestyle for her family of 5 on $24K per year.

1 Want vs. need. Pray for discernment between "do we need it" and "do we just want it." God says he will provide all our needs. So, he will help us live without our wants as well.

2 Plan and save. If you will need it or if you really, really want it, then start saving for it. Paying cash instead of interest puts you dollars ahead. It also helps you avoid impulse purchases and keeping-up-with-the-Joneses buying. Or put off the decision for one month, and see if it's as important to you (or your kids) then.

3 Buy local. Compare the prices at local businesses to the discount places. You may find there is no significant difference. Don't forget to figure in the gas money to drive to the discount place. Our local grocery store is very comparable in prices. They will let me buy by the case. They will order non-stocked items. They provide jobs for my friends. Plus, who wants to drive 60 miles to get a gallon of milk because you didn't support your local store and keep them in business?

4 Know what you need. Make a list of your family's sizes, measurements, and future needs (i.e., sports shoes). Put the list in your purse or wallet. Then when you stumble upon a good deal, you'll know if it will fit. This also keeps Dad and the kids out of trouble when buying for Mom. She won't be insulted with the size they choose.

5 Buy previously owned. Garage sales, secondhand stores, and auctions are great. Have your list (see #4) and a tape measure with you. Two summers ago, we bought our daughter a 25¢ bike. It lasted more than a year. Now it needs tires, but she's grown and needs a bigger bike. We got our 13-year-old a "like new" 15-speed adult size bike for $20. We have purchased an entertainment center, tools, home decor, hunting items, and much more this way. Note: To avoid getting caught up in the bidding at an auction, know your top bid. Stop bidding if it exceeds that.

6 Buy at sales and hide things. Say, for example, you have a hunter on your list, and you find camouflage coats on sale. Buy one, hide it, and there's a birthday gift. Have a drawer, box, or closet to store items bought on sale for bridal, wedding, baby, birthday or whatever gifts. This will save gas, money, and time when go to the drawer for the gift instead of the store.

7 Help each other. The best wheat we produce is usually planted on or near September 23. So James plants the wheat, and I fill and haul the fertilizer and seed to him. This saves James days, miles of walking, and also gives our crop the best planting date possible. James helps pick the garden, snaps beans, and hangs laundry. Help the neighbor work his calves, and he'll help you work yours. Clean out your neighbor's basement, and she'll help you with yours. Oftentimes much more can be done in a day if two or more people do it together.

8 Cut food costs. Canning, buying bulk, and making from scratch are great ways to save money. But what do you do when Mom doesn't have time (because she's doing #7)? When you do have time, try these ideas. Make cookie dough and freeze in cookie-size balls. When frozen place in bags. Then you can thaw and bake as many as you need. Or bake several dozen and freeze. This can also be done with breads. When you make one casserole make three (or more), double foil, label, and freeze. This takes little more time. When you make runzas, bake 5 dozen instead of one, and fill some of them with pizza fixings or broccoli, ham, and cheese. Freeze the extra. Reheat in the oven or microwave.

Get together with a friend whose family likes the same food as yours, and make as many meals as you can, in a day, to put into your freezers. Here is a favorite: Brown 20 pounds of hamburger, sausage, or Italian sausage or bake 6 or 8 chickens or roasts. Debone if necessary, then bag and freeze the cooked meat in meal size packages. These can be turned into sandwiches, omelettes, casseroles, tacos, or spaghetti in just a few minutes. And it is so much cheaper than buying ready-made convenience foods.

9 Be smart about the laundry. Use your clothesline as much as the weather allows. Do the highly advertised brands really clean better? Use half a dryer sheet per load. Save the used dryer sheets and use two in the next load. When you have a burned pan to wash, put a couple of used dryer sheets in it, add very hot water, soak until the water is warm, and then scrub the pan clean.

10 Remember, value isn't always measured in $$. Buy quality so supplies last longer. Kids running a dishwasher may add lots of soap, so the premeasured tabs may be more cost effective. Do you use all of your cell phone minutes each month, or call all 50 states? Look at your records and see how many minutes you used and where you called. Find a plan that better matches your calling activity.

By Lana Skelton

Can you beat this family for frugality?

James and Lana Skelton of Wauneta, Nebraska, and their three children – Josh, 17; Kelsie, 10; and Jesse, 13 – manage a solid middle-class lifestyle on a living allowance of about $24,000 a year. They know, because James and Lana belong to the Nebraska Farm Business Association and they track every penny going in and out of their 3,750-acre corn, wheat, and cattle farm.

Pinching pennies

They pay cash for used cars. Lana cans her garden produce and buys groceries in bulk – rice in 20-pound bags, yeast in 5-pound packages. And she outfits the family in an amazingly stylish garage-sale wardrobe. To be ready for clothing bargains, she says, "This year I took $400 out of the bank and put it in a separate spot in my wallet." She spent $300 on clothing.

There's no sense of deprivation here, though. Josh has been to Tijuana, Mexico, three times with his Hamlet Union Church to build houses. After a week's stay, he usually gives some clothes to poor families.

The children know the value of a dollar. Recently, the boys tried to talk their dad into getting a high-speed Internet connection if they bought a new modem with the money they made selling sweet corn. "So many kids nowadays just need to come out of the big city and spend a week at my farm," James says proudly.

Avoiding false economy

James farms frugally, using older equipment. He dickers with local chemical sellers for a better price.

But he won't buy no-name corn seed from discounters because he can't return the seed if he decides he needs a shorter-season variety during planting. "It's my baby once it rolls out their door," he says. ■

Lana and James Skelton and their children, Kelsie, Jesse, and Josh, dress well with bargain clothes. They buy jeans new. Lana buys bulk groceries but prefers to shop locally.

Outtakes
from the "?" files

A few of the "unusual" *All Around the Farm* submissions *Successful Farming* magazine has received over the years.

Stay cool as a cucumber in a rhubarb patch? One reader claimed you could beat the heat by draping a rhubarb leaf over your head, then topping it off with a cap. We would suggest, however, that you remove the leaves before making that run into town for parts.

One young reader sent in his own illustration of some exercise equipment for his family's hogs. It consists of a semi-trailer truck tire hanging by a rope from the roof of the hog barn.

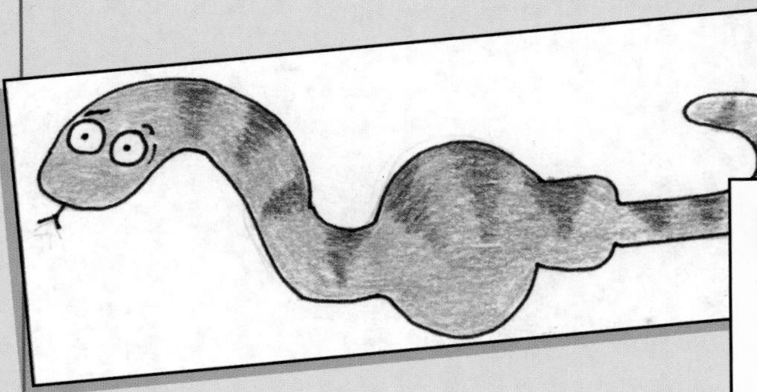

Near-sighted snakes beware. Egg-eating snakes in the henhouse? A reader advised placing white doorknobs in the hens' nests. The snakes swallow the faux eggs, slither off to wherever snakes go, and die. At least that's the theory.

Don't try this with your ATM card. A reader wrote: "We put in keyless deadbolts with a secret code on our shop door and machinery shed. Since the digital code program runs from 2 to 7 digits, memory can get hazy on a frozen morn. Here's what we did." What they did was tape the code to the lock. It worked but so much for the secret code.

Let that be a lesson to you. One solution to a reader's gopher problem, involved harvesting roadkill and stuffing the carcasses down the gopher holes, presumably sending the pesky critters a message they couldn't refuse.

It may look funny, but it's effective. This reader claimed to have circumvented a killing frost by putting lidless and bottomless coffee cans over her recently planted tomatoes and then placing her husbands old caps over the openings. While it kept the tomatoes cozy, it proved to be dangerous, as one passing driver wound up in the ditch while gawking at the colorful rows.

Ever consider chopping alfalfa with a surplus military helicopter? Probably not. But one reader has. Our submitter estimated, we assume tongue-in-cheek, that his Cobra "chopter" could cut "40 acres per hour with practice." A lot of practice.

An Army National Guard Corba attack helicopter sits idle in an alfalfa field north of Diamond Bluff after its pilots landed safely following a mechanical malfunc- ti... hoto by Bill Pond

...rd helicopter forced to ...n Diamond Bluff field

...r end of training ...ithout injuries

were flying at 4,000 feet.

"It's a fairly conservative approach. We're trying to do the safest thing we can," Schmitz said. "It's nothing that

Foam-markers are common in the field, but what if you need to mark a straight line in the yard? Just attach an angle iron with levers to the back of a spreader. Fit cans of shaving cream so that they touch the levers, which are operated by ropes. And if you happen to be growing a beard, you'll never miss the shaving cream.

All around® the farm *A page written by our readers*

We find a floor for a slat cribbing type of round corncrib is easily made by laying concrete building blocks over a circle to form a smooth platform for easy scooping. We leave the center row of blocks out for drag to go in for shelling corn.—*E. W. S., Iowa.*

After we purchased a metal wagon, we found the floor dangerously slippery, especially at haying time. By painting the wagon floor with a thin coat of roofing tar and sifting sand on it, this fault was corrected.—*R. O., Michigan.*

To replace washers in a hydraulic cylinder, I use a piston ring compressor to draw down the washer to the same size as the cylinder. Thus, edges of the washer are not damaged when inserted into the cylinder.—*L. G., Minnesota.*

For doorways of my grain bins, I cut the boards at an angle so I don't always have to push them all the way

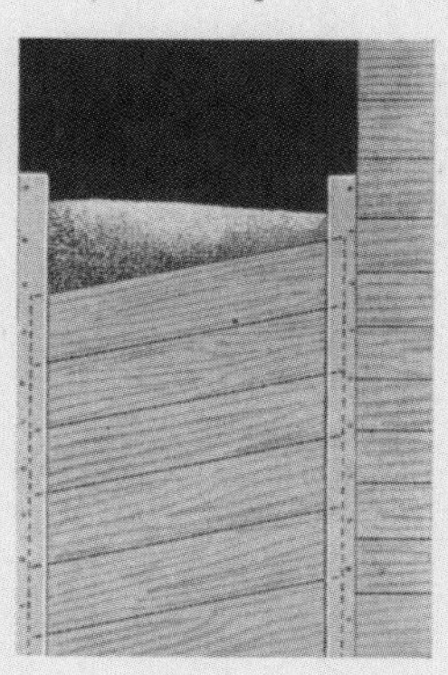

up to the top to remove. I just lift it up on one end, and the board comes out. —*V. M., Ohio.*

On my hydraulic dump wagons, I have mounted two broom holders to hold the hose when not in use. I also

use an ordinary door spring attached to the wagon and a broom holder at the other end to hold the hose if it is too long when in use.—*L. W., Iowa.*

To line my concrete feed mangers, I split lengthwise a glazed tile 24″ in diameter. I set these in concrete to form the manger trough. They provide a very smooth surface and are easily cleaned.—*W. F., Nebraska.*

I put a grab hook on the end of a chain about a foot long with an iron ring on the other end of the chain just big enough for a clevis to pass through. With this on the tractor drawbar, I can grab onto a tow chain at any point more easily, and I don't have to wrap the chain around the drawbar to get it short enough for good towing.—*J. H., Minnesota.*

Each fall after the barn loft is filled, I replace the old light bulbs with new ones and put the old ones to use wherever needed, but in easy-to-reach places. Since doing this I have avoided about all bulb replacement at seasons when it would be a long reach to the top of the near-empty hayloft.—*R. W., Wisconsin.*

For accurate and quick measurement of our fields, we made the measuring wheel as shown. The wheel rim is a 3/8″ round iron bar 10′ long, bent to a circle. Two 38″ lengths of 3/8″ rod are bent to form the spokes, and are welded to the axle member. For the push-handle, two rods of 3/8″ iron 38″ long are set one on each side of the wheel and spread at the handle end by

a 14″ length of 1″ pipe. A baler tally and trip are mounted on a bracket attached to one of the handle rods. A stud welded to a spoke operates the tally or counter trip every time the wheel makes one revolution. Thus, on any run the figure on the baler tally times 10 is the measurement in feet. The counter is then reset, and the next run is made.—*J. B. C., Indiana.*

To make a striker board for laying walks and so on, we fasten together two straight boards in the center as shown. These boards are spread apart at each end, and a heavy spacer holds them securely. The spacer serves also as a handle. The spread of the boards at the end resting on the forms keeps the striker standing vertically on edge. The curve of the bow also tends to keep the concrete running toward the center instead of spilling over sides of forms.—*A. O., Colorado.*

Drawings: Associated Artists

Successful Farming invites you to share your ideas with the rest of our readers. We pay $5 for each item upon publication, but regret we cannot return those we don't use.—*The Editors*

Keeps Shop Orderly

ORDER can be established in the farm shop by having handy places to keep small supplies such as different sizes of nails and screws, washers, stove bolts, rivets, and similar items. The buckets from an old elevator belt when nailed or screwed to the wall in an orderly fashion are just the thing. The name of each article can be painted on the bucket as a further help in maintaining order.—D. M., Ill.

Strengthens Doubletree

A doubletree can be made stronger with only a little more work and expense. Instead of boring a hole thru the center of it as is usually done two pieces of iron are shaped in a forge to carry the draw pin, as is shown in the drawing. This arrangement saves the wear which would take place in a

hole bored into the wood and the entire width of the doubletree withstands the pull. A notch is cut to fit the iron and they are fastened on with lag screws. —H. E. Ill.

For Trimming Shelves

Crepe paper can be given an attractive scalloped edge very easily. Cut out a strip of the paper of the desired width. Lay it on a smooth surface and place the first and second fingers of the left hand three-fourths of an inch apart near one edge of the strip. Place the forefinger of the other hand between these two fingers and on the edge of the paper. Press down firmly and pull that finger back across the strip. This leaves a scallop. Repeat the entire length of the strip. This makes a neat decoration for cupboard shelves and it can be used in many other places, too.—Mrs. E. A. S., Ind.

Hay Stays in Manger

Keeping cows from tossing the hay out of their mangers is often a problem. We solved it by stretching a wire along the manger top. Erect a 2 x 4 solidly at each end of the manger so that their centers come about five inches back of the inner top edge of the manger and extend about ten inches higher than the top of the manger. A hole is drilled in each one about three inches from the top and eyebolts inserted. A No. 9 wire is stretched between them. The wire is connected with a coil spring to the eyebolt at one end which keeps the wire tight but allows it to give if bumped or hooked by a cow. Slack can be taken up by tightening the bolts. This wire effectively stops the hay tossed up by the cows.—F. G. S., Minn.

If you have some clever, original device for lightening the farm or home work, saving time, advertising or selling the products of the farm or improving home surroundings, tell us about it. Give all necessary details, but be brief. Send picture or drawing if possible. For each idea published in this department we will pay $2. It is impossible to return unused suggestions.

Makes Square Turn Possible

A simple attachment to the tractor drawbar for hooking up the binder has eliminated split binder tongues for us. Attaching directly to the drawbar as shown under "the old way" resulted in a split tongue if a square turn is made

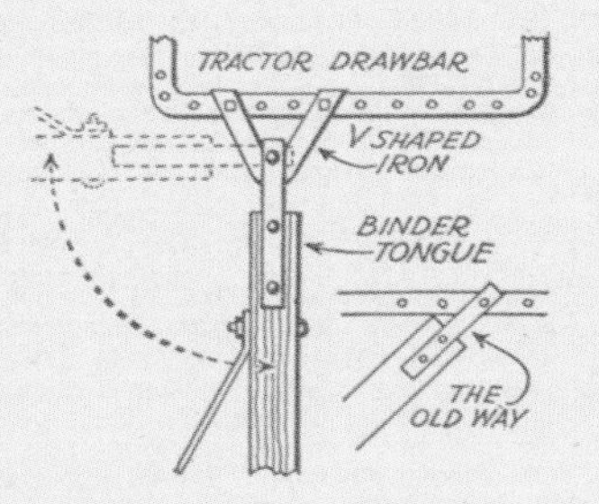

at the corners. The V-shaped piece makes possible a sharp, right-angled turn and a neater job of cutting the field as well as preventing delays which a broken tongue would cause. The attachment is made of iron a little lighter than used in the drawbar.—K. W., Iowa.

Holds Troughs Together

Here is a way to secure the ends of hog troughs so they will stay in place. In addition to the nails ordinarily used to fasten the end pieces drive some long nails thru the end pieces so as to miss the ends of the trough to project on inside. Then take pliers and bend a right-angled hook on each nail as seen in the illustration and drive them in the wood. Do the same thing on the underside.—P. E. B., Iowa.

A Better Barbed Wire Holder

The common method of unrolling barbed wire by putting an iron bar thru the spool can be greatly improved upon. Drill a hole thru this bar about one inch out from each edge of the spool; put discarded discs from a disc harrow over the rod, concave side out; put bolts thru the holes to hold the discs on and it will be much more convenient to handle.—R. A., Iowa.

Use the Old Oil

I find that the most efficient way to oil a plow or cultivator is to use old oil, heavy if possible, and apply with an old paint brush. This does a good job and it is not necessary to get oil on the hands, either. This system is especially good in oiling the sickle on the mower. It is about the easiest way to oil the springs on a car for which a light oil should be used, separator oil preferred. —A. T. L., Iowa.

Cleans Light Shoes

To clean your light shoes, just rub lightly with a piece of art gum. All surface dirt will be removed as if by magic. I always carry a piece of this in my purse, as it is invaluable in keeping up a neat appearance while away from home.—J. C., Ill.

Bottles Stay on Shelf

To prevent bottles from falling off a shelf, tack a rubber suction cup to the wall behind the position for each bottle and push the bottle against it with slight pressure. When needed the bottle can be removed, but it will not fall off the shelf from a little bump or jar.—C. T. S., Wis.

Wire Brush Saves Time

A small inexpensive wire bristle brush can be used to advantage around the home. It may be used to remove the lint and threads from the brush of either a hand or electric sweeper, for the lint can be burned from the wire brush without injuring the brush. It is useful, also, in removing paint. It comes in handy in cleaning caked dirt and oil from automobile springs prior to oiling.—F. B. B., Pa.

Handy Feed Scoop

A discarded flour sifter proved to be just the thing for converting into a handy grain scoop. All the fixing it needed was the removal of the revolving sifter part. It holds just a quart which makes it convenient in measuring the right amount of feed for each cow.—G. R., Mich.

APPROXIMATE CONVERSIONS

When You Know:	You Can Find	If You Multiply By:
inches	millimeters	25
feet	centimeters	30
yards	meters	0.9
miles	kilometers	1.6
millimeters	inches	0.04
centimeters	Inches	0.4
meters	yards	1.1
kilometers	miles	0.6
ounces	grams	28
pounds	kilograms	0.45
short tons	megagrams (metric tons)	0.9
grams	ounces	0.035
kilograms	pounds	2.2
megagrams (metric tons)	short tons	1.1
fluid ounces	milliliters	30
pints	liters	0.47
quarts	liters	0.95
gallons	liters	3.8
cubic inches	cubic centimeters	16.4
milliliters	fluid ounces	0.034
liters	pints	2.1
liters	quarts	1.06
liters	gallons	0.26
cubic centimeters	cubic inches	0.06
sq. inches	sq. centimeters	6.5
sq. feet	sq. meters	0.09
sq. yards	sq. meters	0.8
sq. miles	sq. kilometers	2.6
acres	sq. hectometers (hectares)	0.4
sq. centimeters	sq. inches	0.16
sq. meters	sq. yards	1.2
sq. kilometers	sq. miles	0.4
sq. hectometers (hectares)	acres	2.5

CUBIC MEASURE

1728 cubic inches	1 cu. foot
27 cubic feet	1 cu. yard
128 cubic feet (8'×4'×4')	1 cord
1'×1'×1'	1 bd ft

WEIGHT MEASURE

gram	15.432 grains
gram	.0353 ounce
kilogram	2.2046 pounds
kilogram	.0011 ton (short)
metric ton	1.1025 ton (short)
grain	.064 grams
ounce	28.35 grams
16 ounces	1 pound
1 pound	453.5 grams
100 pounds	1 hundredweight
20 hundredweight	1 ton
ton (short)	907.18 kilograms
ton (short)	2,000 pounds
ton (long)	2,240 pounds
1 gallon of water	8.34 pounds

AVOIRDUPOIS WEIGHT

27 11/32 grains	1 dram
16 drams	1 ounce
16 ounces	1 pound
100 pounds	1 hundredweight
2000 pounds	1 ton
2240 pounds	1 gross or long ton

LINEAR MEASURE

12 inches	1 foot
3 feet	1 yard
5½ yards (16½ ft.)	1 rod
320 rods	1 mile
1760 yards (5,280 ft.)	1 mile

SQUARE MEASURE

144 square inches	1 square foot
9 square feet	1 square yard
30¼ square yards	1 square rod
160 square rods	1 acre (43,560 sq ft)
640 acres	1 square mile
1 square mile	1 section

LIQUID/DRY MEASURE

16 fluid ounces or 1 pound (water)	1 pint
2 pints	1 quart
4 quarts	1 gallon
8 quarts	1 peck
4 pecks	1 bushel
1 bushel	1¼ cu. ft.
2 barrels	1 hogshead
31½ gallons (US)	1 barrel
7.48 gallons (US)	1 cu. ft.
1 gallon (US)	231 cu. in.
1 inch of water per acre	27,154 gallons
1 inch of water per hectare (1 hectare = 2.5 acres)	67,885 gallons

TRAVEL SPEED CONVERSIONS

Most travel speeds are for tractors with 18.4-38 R-1 rear tires. Adjust as follows for other tires.

Tire	Type	Adjustment
15.5 - 38	R-1	9% slower
16.9 - 38	R-1	3% slower
18.4 - 34	R-1	7% slower
18.4 - 38	R-2	1% faster
18.4 - 42	R-1	7% faster
20.8 - 34	R-1	4% slower
20.8 - 38	R-1	4% faster
20.8 - 38	R-2	6% faster
20.8 - 42	R-1	10% faster
23.1 - 34	R-1	1% faster
23.1 - 34	R-2	6% faster

R1 TYPE - REGULAR TREAD, DRIVE WHEEL

Tire Load Limits (lbs.) at various cold inflation pressures (PSI)

Tire Size Designation	18★	24★★	30★★★
18.4R38	6000*	7150**	
18.4R42	6400*	7400**	8550***
20.8R38	7150*	8550**	
20.8R42	7600*	9100**	

NOTE:★★★(3-star) radial not recommended until further notice. 3 star tires require special rims.

PASSENGER CAR & LIGHT TRUCK TIRE ROTATION PATTERNS

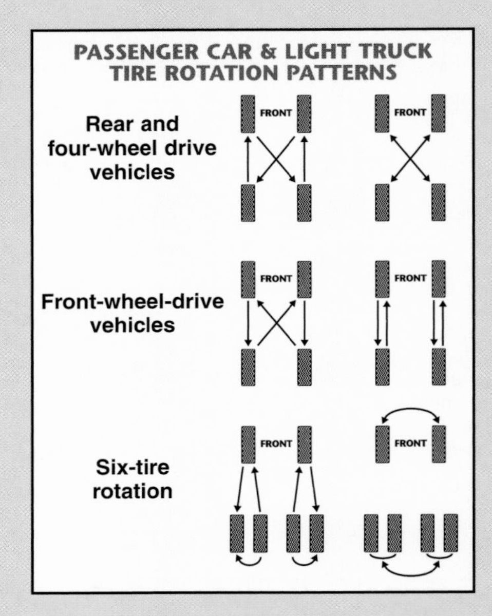

THREE-PHASE A.C. MOTOR CURRENTS

	208 Volts				230 Volts		460 Volts	
	Use Current Values Below For Determining Minimum Conductor Ampacity*		Use Current Values Below For Selecting Conductors Using 230-240 Volt Tables For Voltage Drop**					
Motor Horse-power	Full Load (Amps.) Col. 1	125% Full Load (Amps.) Col. 2	Voltage Drop Full Load (Amps.) Col. 3	Voltage Drop 125% Full Load (Amps.) Col. 4	Full Load (Amps.) Col. 5	125% Full Load (Amps.) Col. 6	Full Load (Amps.) Col. 7	125% Full Load (Amps.) Col. 8
½	2.2	2.8	2.5	3.2	2.0	2.5	1.8	2.2
¾	3.1	3.9	3.6	4.4	2.8	3.5	2.6	3.2
1	4.0	5.0	4.6	5.7	3.6	4.5	3.4	4.2
1½	5.7	7.1	6.6	8.2	5.2	6.5	4.8	6.0
2	7.5	9.4	8.6	11	6.8	8.5	7.6	9.5
3	11	14	12	15	9.6	12	11	13.8
5	17	21	19	24	15.2	19	14	18
7½	24	30	28	35	22	28	21	26
10	31	39	36	44	28	35	27	34
15	46	58	53	67	42	52	34	42
20	59	74	68	86	54	68	40	50
25	75	94	86	108	68	85	52	65
30	88	110	102	128	80	100	65	81
40	114	142	132	165	104	130	77	96
50	143	179	165	206	130	162	96	120
60	169	211	195	244	154	192	124	155
75	211	264	244	305	192	240	156	195
100	273	341	315	394	248	310	180	225
125	343	429	396	495	312	390	240	300
150					360			
200					480			

*Use these values of current **only** for determining the minimum conductor ampacity.

**Use values in Cols. 3 and 4 for selecting conductor size in relation to voltage drop. If motor current exceeds value in Col. 1, determine values for Col. 3 by multiplying motor current by 1.15, and values for Col. 4 by multiplying motor current by 1.44.

WIRE INSULATION CLASSIFICATIONS

Insulation Material	Type Letter	Description
Rubber	RHW	Moisture-and heat-resistant rubber
Latex rubber	RUW	Moisture-resistant latex rubber
Thermoplastic	TW	Moisture-resistant thermoplastic
	THW	Moisture-and heat-resistant thermoplastic
	THWN	Flame retardant, moisture- and heat-resistant, with nylon jacket outer covering
Cross-linked synthetic polymer	XHHW	Flame retardant
Hard service cord	SO	Thermoset insulated with oil-resistant thermoset cover, no fabric braid
	STO	Thermoplastic or thermoset insulated with oil-resistant thermoplastic cover, no fabric braid

EFFECT OF VOLTAGE DROP ON POWER AND LIGHT LOSS

Voltage drop %	Light loss %	Power loss %
1	3	1.5
2	7	3.0
3	10	4.5
4	13	6.0
5	16	7.5
10	31	15.0
15	46	22.5

MAXIMUM NUMBER OF WIRES IN PVC CONDUIT

AWG= American Wire Gauge

Insulation type letter	Wire Size AWG	Conduit Size ½"	¾"	1"
RHW (with outer covering)	14	3	6	10
	12	3	5	9
	10	2	4	7
	8	1	2	4
THWN	14	13	24	39
	12	10	18	29
	10	6	11	18
	8	3	5	9
TW, RUW, XHHW	14	9	15	25
	12	7	12	19
	10	5	9	15
	8	2	4	7
THW, RHW (without outer covering)	14	6	10	16
	12	4	8	13
	10	4	6	11
	8	1	3	5

SINGLE-PHASE A.C. MOTOR CURRENTS

Motor Horse-power	115 Volts Full Load (Amps.)	115 Volts 125% Full Load (Amps.)	230 Volts Full Load (Amps.)	230 Volts 125% Full Load (Amps.)
⅛	4.4	5.5	2.2	2.8
¼	5.8	7.2	2.9	3.6
⅓	7.2	9.0	3.6	4.5
½	9.8	12.2	4.9	6.1
¾	13.8	17.2	6.9	8.6
1	16	20	8.0	10
1½	20	25	10.0	12.5
2	24	30	12	15
3	34	42	17	21
5	56	70	28	35
7½	80	-	40	50
10	100	-	50	62

AMPACITY OF COPPER WIRE

Column A = **TW** & **RUW** wire **UF** & **NM-B cable**.

Column B = **THW**, **THWN**, **RHW** & **XHW** wire & **USE** cable.

AWG 14-CU wire is only used for some equipment - not for branch circuits.

Wire Size AWG-CU	Ampacity A	amperes B
14	15	15
12	20	20
10	30	30
8	50	40
6	55	65
4	70	85
3	85	100
2	95	115
1	110	130
0	125	150
00	145	175
000	165	200
0000	195	230

DUCT AREAS FOR VARIOUS SIZE DIAMETERS OF AERATION DUCTS

Duct Dia.	Cross Sectional Duct Area (Sq. Ft.)	Effective Surface Area Per Foot of Length (Sq. Ft)*
12"	.785	2.51
14"	1.07	2.93
16"	1.40	3.35
20"	2.18	4.18
24"	3.14	5.02

*80% of Actual Surface Area

CONCRETE MIXES

	Max. size aggregate	Gallons of water for each sack of cement, using:[1] Damp[3] sand	Wet[4] (average) sand	Very[5] wet sand	Suggested mixture for 1-sack trial batches:[2] Cement sacks cu. ft.	Fine cu. ft.	Aggregate Coarse cu. ft.
		- gallons -			- cubic ft. -		
5-gallon mix: use for concrete subjected to severe wear, weather, or weak acid and alkali solutions.	¾"	4½	4	3½	1	2	2¼
6-gallon mix: use for floors (home, barn), driveways, walks, septic tanks, storage tanks, structural concrete.	8 1½"	5½	5	4½	1	2½	3½
7-gallon mix: use for foundation walls, footings, mass concrete.	1½"	6¼	5½	4¾	1	3	4

Make a trial batch to check for slump and workability.
[1]Increase the proportion of water to cement reduces the strength and durability of concrete. Adjust the proportions of trial batches without changing the water-cement ratio. Reduce gravel to improve smoothness; reduce both sand and gravel to reduce stiffness.
[2]Proportions vary slightly depending on gradation of aggregates.
[3]Damp sand falls apart after being squeezed in the palm of the hand.
[4]Wet sand balls in the hand when squeezed, but leaves no moisture on the palm.
[5]Very wet sand has been recently rained on or pumped.

AREAS COVERED BY ONE CUBIC YARD OF MIXED CONCRETE

Depth, in.	Sq. ft.	Depth, in.	Sq. ft.	Depth, in.	Sq. ft.
1	324	4¾	68	8½	38
1¼	259	5	65	8¾	37
1½	216	5¼	62	9	36
1¾	185	5½	59	9¼	35
2	162	5¾	56	9½	34
2¼	144	6	54	9¾	33
2½	130	6¼	52	10	32½
2¾	118	6½	50	10¼	31½
3	108	6¾	48	10½	31
3¼	100	7	46	10¾	30
3½	93	7¼	45	11	29½
3¾	86	7½	43	11¼	29
4	81	7¾	42	11½	28
4¼	76	8	40	11¾	27½
4½	72	8¼	39	12	27

RECOMMENDED UNREINFORCED SLAB THICKNESS

Use	Thickness	Relative strength 4" = 100%
½ ton pickup or less	4"	100%
Small trucks, most farm machinery, & wagons; occasional larger trucks, moderate-sized loaders, manure tanks or spreaders	5"	145%
Frequent large trucks, grain wagons, or manure tanks	8"	400%

*Thickness depends on both size and frequency of maximum load.

DURABILITY AND STRENGTH FOR AIR-ENTRAINED CONCRETE

Durability (weathering and chemical resistance) and strength depend primarily on water-cement ratio. Adding extra water rapidly lowers durability and strength. Only ½ gal/bag (about 3 gal/yd) separates the groups in the table.

These are approximate guidelines - follow plans or specifications if available.

Kind of Job	Approx. strength PSI	Gallons water/bag cement	Water-cement ratio, lb. water per lb. cement
Feedbunks, slats, above ground banker silos	4,500	5.0	.44
Unventilated manure tanks, parking lots, underground silos	4,000	5.5	.49
Feedlots, floors, walls, drives, basements; open top or ventilated manure tanks, reinforced retaining walls, beams and columns	3,500	6.0	.53
Footings, foundation walls, gravity retaining walls	3,000	6.5	.62

RECOMMENDED MORTAR MIXES BY VOLUME

Cement and lime sacks contain 1 cu. ft.
Masonry cement is ASTM Specification C91 Type Type II.
About 6 cu. ft. of mortar will lay 100 standard 8" blocks.

Type & Service	Cement	Hydrated lime	Mortar sand in damp, loose condition
N: Ordinary Service	1-masonry cement	-	2¼ to 3
	or		
	1-portland cement	½ to 1¼	4½ to 6
M: Heavy loads or frost	1-masonry cement plus 1-portland cement	-	4½ to 6
	or		
	1-portland cement	¼	2¼ to 3

WOOD SCREW INFORMATION

Wood Screw Size	Shank Hole Size	Pilot Hole Size		Auger Bit Sizes By 16th For Counterbore Hole
	All Woods	Hardwood	Softwood	
	Nearest Fractional Size Drill	Nearest Fractional Size Drill	Nearest Fractional Size Drill	
0	1/16	1/32	-	-
1	5/64	1/32	1/32	-
2	3/32	3/64	1/32	3
3	7/64	1/16	3/64	4
4	7/64	1/16	3/64	4
5	1/8	5/64	1/16	4
6	9/64	5/64	1/16	5
7	5/32	3/32	1/16	5
8	11/64	3/32	5/64	6
9	3/16	7/64	5/64	6
10	3/16	7/64	3/32	6
11	13/64	1/8	3/32	7
12	7/32	1/8	7/64	7

Standard screw lengths in inches are 3/16, ¼, ⅜, ½, ⅝, ¾, ⅞, 1, 1¼, 1½, 1¾, 2, 2¼, 2½, 2¾, 3, 3½ and up.

BOARD FEET OF LUMBER IN VARIOUS SIZES OF BOARDS

Dimension or size	Length of boards						
	8 feet	10 feet	12 feet	14 feet	16 feet	18 feet	20 feet
1×4	2⅔	3⅓	4	4⅔	5⅓	6	6⅔
1×6	4	5	6	7	8	9	10
1×8	5⅓	6⅔	8	9⅓	10⅔	12	13⅓
1×10	6⅔	8⅓	10	11⅔	13⅓	15	16⅔
1×12	8	10	12	14	16	18	20
2×4	5⅓	6⅔	8	9⅓	10⅔	12	13⅓
2×6	8	10	12	14	16	18	20
2×8	10⅔	13⅓	16	18⅔	21⅓	24	26⅔
2×10	13⅓	16⅔	20	23⅓	26⅔	30	33⅓
2×12	16	20	24	28	32	36	40
4×4	10⅔	13⅓	16	18⅔	21⅓	24	26⅔
4×6	16	20	21	28	32	36	40

ABRASIVE PAPER GRADES

Grit	Equiv. "O" Series	General Description	Remarks
600	-	Super Fine	Range of abrasive papers used for wet sanding
500	-		
400	10/0		
360	-		
320	9/0		
280	8/0	Very Fine	Used for dry sanding all finishing undercoats.
240	7/0		
220	6/0		For final sanding of bare wood.
180	5/0	Fine	Good for smoothing old paint.
150	4/0		
120	3/0	Medium	Use for General wood sanding. Good for first smoothing of old paint, and plaster patches.
100	2/0		
80	1/0		
60	½	Coarse	For rough-wood sanding.
50	1		
40	1½		
36	2	Very Coarse	Too coarse for pad sanders. Heavy machines and high speed are recommended.
30	2½		
24	3		

NAIL SIZES AND WEIGHTS

Size	Length Inches	Wire Gauge	Approx. No./Lb.
Common Nails			
2d	1	15	847
3d	1¼	14	543
4d	1½	12½	294
5d	1¾	12½	254
6d	2	11½	167
7d	2¼	11½	150
8d	2½	10¼	101
10d	3	9	69
16d	3½	8	49
20d	4	6	31
40d	5	4	18
60d	6	2	11
Spikes			
16d	3½	5	24
20d	4	4	19
40d	5	2	12
60d	6	1	9
Finishing Nails			
3d	1¼	15½	807
4d	1½	15	584
6d	2	13	288
8d	2½	12½	189
10d	3	11½	121
Box Nails			
3d	1¼	14½	588
4d	1½	14	453
5d	1¾	14	389
6d	2	12½	225
7d	2¼	12½	200
8d	2½	11½	136

COMPARISON OF THE INTERNATIONAL METRIC SYSTEM AND THE ENGLISH SYSTEM MEASUREMENT

1 Centimeter	= .3937 inches	1 Kilometer	= 1000 meters	1 Gallon	= 3.785 liters
1 Inch	= 2.54 centimeter	1 Kilometer	= .62137 miles	1 Gram	= 15.43 grains
1 Foot	= 30.48 centimeter	1 Sq. Centimeter	= .155q. inches	1 Ounce	= 28.35 grams
1 Meter	= 39.37 inches	1 Sq. Decimeter	= 100 cu. centimeters	1 Kilogram	= 1000 grams
1 Meter	= 100 centimeters	1 Cu. Centimeter	= .061 cu. inches	1 Kilogram	= 2.205 pounds
1 Meter	= 1.094 yards	1 Cu. Decimeter	= 1000 cu. centimeters	1 Pound	= 7000 grains
1 Meter	= 1000 millimeters	1 Cu. Meter	= 100 liters	1 Pound	= .4536 kilograms
1 Millimeter	= .001 meter	1 Fluid Ounce	= 29.54 milliliters	1 Kilogram	= 1000 milliliters
1 Yard	= .9144 meter	1 Liter	= 1000 cu. centimeters	1 Kilogram	= 1 liter
1 Mile	= 1609.344 meters	1 Liter	= 1.057 quarts		

MINIMUM COPPER CONDUCTOR SIZE FOR BRANCH CIRCUITS

Conductor length is one way. Based on the conductor ampacity or 2% voltage drop, whichever is limiting.

For type TW and RUW wire or UF and NM-B cable.

120 V service — Conductor length

Load A	30'	40'	50'	60'	75'	100'	125'	150'	175'	200'
5	12	12	12	12	12	12	12	10	10	10
7	12	12	12	12	12	12	10	10	8	8
10	12	12	12	12	10	10	8	8	8	6
15	12	12	10	10	10	8	6	6	6	4
20	12	10	10	8	8	6	6	4	4	4
25	10	10	8	8	6	6	4	4	4	3
30	10	8	8	8	6	4	4	4	3	2
35	8	8	8	6	6	4	4	3	2	2
40	8	8	6	6	4	4	3	2	2	1
45	6	6	6	6	4	4	3	2	1	1
50	6	6	6	4	4	3	2	1	1	0
60	4	4	4	4	4	2	1	1	0	00
70	4	4	4	4	3	2	1	0	00	00
80	2	2	2	2	2	1	0	00	00	000
90	2	2	2	2	2	1	0	00	000	000
100	1	1	1	1	1	0	00	000	000	0000

240 V service — Conductor length

Load A	50'	60'	75'	100'	125'	150'	175'	200'	225'	250'
5	12	12	12	12	12	12	12	10	12	12
7	12	12	12	12	12	12	12	10	10	10
10	12	12	12	12	12	10	10	8	10	8
15	12	12	12	10	10	10	8	6	8	6
20	12	12	10	10	8	8	8	6	6	6
25	10	10	10	8	8	6	6	6	6	4
30	10	10	10	8	6	6	6	4	4	4
35	8	8	8	6	6	6	4	4	4	4
40	8	8	8	6	6	4	4	4	4	3
45	6	6	6	6	6	4	4	4	3	3
50	6	6	6	6	4	4	4	3	3	2
60	4	4	4	4	4	4	3	2	2	1
70	4	4	4	4	4	3	2	2	1	1
80	2	2	2	2	2	2	2	1	1	0
90	2	2	2	2	2	2	1	1	0	0
100	1	1	1	1	1	1	1	0	0	00

For type THW, THWN, RHW, and XHHW wire.

120 V service — Conductor length

Load A	30'	40'	50'	60'	75'	100'	125'	150'	175'	200'
5	12	12	12	12	12	12	12	10	10	10
7	12	12	12	12	12	12	10	10	8	8
10	12	12	12	12	10	10	8	8	8	6
15	12	12	10	10	10	8	6	6	6	4
20	12	10	10	8	8	6	6	4	4	4
25	10	10	8	8	6	6	4	4	4	3
30	10	8	8	8	6	4	4	4	3	2
35	8	8	8	6	6	4	4	3	2	2
40	8	8	6	6	4	4	3	2	2	1
45	8	8	6	6	4	4	3	2	1	1
50	6	6	6	4	4	3	2	1	1	0
60	6	6	4	4	4	2	1	1	0	00
70	4	4	4	4	3	2	1	0	00	00
80	4	4	4	3	2	1	0	00	00	000
90	3	3	3	3	2	1	0	00	000	000
100	3	3	3	2	1	0	00	000	000	0000

240 V service — Conductor length

Load A	50'	60'	75'	100'	125'	150'	175'	200'	225'	250'
5	12	12	12	12	12	12	12	12	12	12
7	12	12	12	12	12	12	12	12	10	10
10	12	12	12	12	12	10	10	10	10	8
15	12	12	12	10	10	10	8	8	8	6
20	12	12	10	10	8	8	8	6	6	6
25	10	10	10	8	8	6	6	6	6	4
30	10	10	10	8	6	6	6	4	4	4
35	8	8	8	8	6	6	4	4	4	4
40	8	8	8	6	6	4	4	4	4	3
45	8	8	8	6	6	4	4	4	3	3
50	6	6	6	6	4	4	4	3	3	2
60	6	6	6	4	4	4	3	2	2	1
70	4	4	4	4	4	3	2	2	1	1
80	4	4	4	4	3	2	2	1	1	0
90	3	3	3	3	3	2	1	1	0	0
100	3	3	3	3	2	1	1	0	0	00

RECOMMENDED RPM OF HIGH-SPEED DRILLS IN VARIOUS METALS*					
Drill Size**	Wrought Iron, Low-Carbon Steel	Medium-Carbon Steel	High-Carbon Tool Steel	Cast Iron	Aluminum & Brass
1/16	5000-6700	4000-4900	3000-3600	4200-6100	12,000-18,000
1/8	2500-3350	2100-2450	1500-1800	2100-3000	6000-9000
3/16	1600-2200	1400-1600	1000-1200	1400-2000	4000-6000
1/4	1200-1700	1000-1200	750-900	1100-1500	3000-4500
5/16	1000-1300	850-950	600-725	850-1200	2400-3600
3/8	800-1100	700-800	500-600	700-1000	2000-3000
7/16	700-950	600-700	425-525	600-875	1700-2500
1/2	600-850	450-600	375-450	525-750	1500-2250
9/16	550-750	425-550	340-400	475-675	1325-2000
5/8	500-650	400-500	300-350	425-600	1200-1800
11/16	450-600	375-450	275-325	375-550	1100-1650
3/4	400-550	350-400	250-300	350-500	1000-1500
13/16	375-500	325-375	235-275	325-460	900-1375
7/8	350-450	300-350	225-250	300-425	850-1275
15/16	325-425	275-325	210-235	275-400	800-1200
1	300-400	250-300	200-225	250-375	750-1125

*Reduce rpm one-half for carbon drills.
**For number and letter drills, use speed of nearest fractional-sized drill.

HEIGHT OF TREE OR BUILDING

To estimate the height of a tree or building:

1. Measure the height (H_1) of a nearby object that is vertical.
2. Measure the length of the shadow (S_1) cast by that object.
3. Measure the length of the shadow (S_2) cast by the tree or building.
4. Length of shadow (S_2) of tree or building times the height (H_1) of the object divided by length of shadow (S_1) of the object equals the height (H_2) of the tree or building ($H_2 = S_2 \times H_1 / S_1$).

Test / Metal	Low-carbon steel	Medium-carbon steel	High-carbon steel	High-sulphur steel	Manganese steel	Stainless steel	Cast iron	Wrought iron
APPEARANCE	Dark grey	Dark grey	Dark grey	Dark grey	Dull, cast surface	Bright, silvery smooth	Dull grey; evidence of sand mold	Light grey; smooth
MAGNETIC	Strongly magnetic	Strongly magnetic	Strongly magnetic	Strongly magnetic	Non-magnetic	Depends on exact analysis	Strongly magnetic	Strongly magnetic
CHISEL	Continuous chip; smooth edges; chips easily	Continuous chip; smooth edges; chips easily	Hard to chip; can be continuous	Continuous chip; smooth edges; chips easily	Extremely hard to chisel	Continuous chip; smooth, bright color	Small chips, about 1/8 in.; not easy to chip; brittle	Continuous chip; smooth edges; soft and easily cut and chipped
FRACTURE	Bright grey	Very light grey	Very light grey	Bright grey; fine grain	Coarse grained	Depends on type; bright	Brittle	Bright grey; fibrous appearance
FLAME	Melts fast; becomes bright red before melting	Melts fast; becomes bright red before melting	Melts fast; becomes bright red before melting	Melts fast; becomes bright red before melting	Melts fast; becomes bright red before melting	Melts fast; becomes bright red before melting	Melts slowly; becomes dull red before melting	Melts fast; becomes bright red before melting

ELECTRODES FOR WELDING

Type of Work	Electrode AWS No.	Electrode NEMA Color	Current & Polarity (S-straight) (R-reversed)	Welding Position	Penetration	Characteristics	Uses
MILD STEEL	E6010 E6011	None Blue spot	DC-R All	All All	Deep Deep	Digging affects it; leaves rough, rippled surface; slow burn off; a strong weld with much spatter.	All mild-steel welding; 1/8" very good for holes and cutting; best farm electrode.
	E6012	White spot	All	All All	Medium	Melted metal is gummy.	Fill in gaps and poor fit ups; medium arc keeps sagging metal from closing the circuit.
	E6013	Brown spot	All	All	Shallow		Very good for down welds on thin sheet metal.
	E6013	Brown spot	All	All	Shallow	General-purpose; easy to use; less burn-through owing to shallower penetration.	All mild steel; for extra strength, lay a heavy first bead or two light beads.
	E6014	Brown spot	All	All	Shallow	Cross between E6013 and E6024; easy to use; fast burn off.	Second best for farm; same uses as above
	E6024	Yellow spot	All	Flat, Horizontal	Medium	Fast deposit; slag removes itself if amperage is right and edges not pinned by poor electrode motion.	All downhand welds, in fillets and horizontal; 3/32" good as spray rod on vertical down welds.
	E6027	Silver spot	Flat, horizontal	All	Medium	Less undercut than E6024; better weld; slag not as easy removed; less penetration on rusty metal.	Same as E6024, but see Characteristics (left).
LOW-ALLOY METAL	E7016	See Character	AC, DC-R	All	Varies	Requires clean surface*; NEMA code: green group, orange spot, blue end.	All difficult welds; rather hard for beginners to use.
NON-MACHINABLE CAST IRON	No AWS; non-machinable cast iron	Orange end	AC, DC-S	All	Medium	Requires clean surface*; low heat needed reduces cracking of weld or work: *Do not permit work to become dark red.*	For shielded arc welds if not machined; hold close arc, but coating must not touch molten metal; intermittent beads not over 3" long; peen lightly after each bead; cool and clean before next bead.
CAST IRON	No AWS; cast iron	Orange spot	AC, DC-R	All	Medium	Same as above, except machinable.	All cast iron, machinery gears, housings, parts, mold boards, etc.
STAINLESS STEEL	No AWS; s.steel 308-16	Yellow group & end	All	All	Medium	Requires short arc, electrode held 15° in direction of travel.	High-to-medium carbon-alloy steel and most nonferrous metals; auto bumpers, mold boards, etc.
HARD-SURFACING	No AWS numbers or NEMA color. Consult manufacturers' catalogs according to uses.		AC, DC-R	All	Light	For all hardsurfacing: requires clean surface*; not for joining parts; build up worn parts with a strength rod, then hardsurface; for wear resistance, lay straight or weavy bead not over 3/4" wide; remove slag before new bead.	Metal-to-metal wear: building-up gives moderate hardness to resist shock and abrasion.
			AC, DC-S	All	Light		Metal in rocky soil: build-up for resistance to impact and severe abrasion; for mild or carbon steel, low-alloy, or high-manganese steel.
			AC, DC-S	All	Light		Metal in sandy soil: to resist any kind of abrasion, mild impact; for carbon, alloy, or manganese steel.
			AC, AC torch	All	Light		Knife edges: fine-grain alloy powder applied with carbon arc gives smooth, abrasion-resistance surface.

*This means that surfaces must be cleaned of rust, grease, oil, moisture, and other foreign matter by wire brushing or grinding.